AI CONVERSATIONS THAT WORK

AN ETHICAL PROFESSIONAL'S GUIDE TO DIRECTING ARTIFICIAL INTELLIGENCE

JOSHUA M. DOTSON

WRALE LTD

AI Conversations That Work

An Ethical Professional's Guide to Directing Artificial Intelligence

First Edition: October 2024

ISBN: 978-1-966151-02-9

Published by: Wrale Ltd, wrale.com

Author Contact: contact@jmdots.com

CONTENTS

FOREWORD

Talking to AI is generally more of an art than a science. My realization of that is the origin of the book you're holding.

My experience in AI communication began with fumbling hobby experiments that led to surprising discoveries. I quickly learned that the most valuable insights often come from hands-on experience, from those moments when AI doesn't entirely give you what you expect. I learn best when I find ways around problems I encounter.

This isn't a textbook packed with math formulas. It's a practical guide drawn from my direct experience with AI and what I've discovered works. I'll show you how to think differently about your interactions with AI and amplify your creative output with their help.

I believe in the power of firsthand experience. As you read this book, experiment with AI in parallel. Try out the concepts, test the methods, and don't be afraid to stumble. The challenges you encounter will make the solutions offered here all the more meaningful.

Remember, every AI conversation is an opportunity to learn and refine your approach. The art of AI communication is rapidly evolving, “and with this book, I hope to provide you with a canvas and some initial brushstrokes. The projects you create are uniquely your own.

Welcome to the fascinating world of AI conversational techniques. Let’s get on with the action.

– Joshua,

Late October 2024

PREFACE

Communicating well with AI is an essential skill. This book, "AI Conversations That Work: An Ethical Professional's Guide to Directing Artificial Intelligence," is a fresh take on AI communication techniques.

Purpose of the Book

The primary purpose of this book is to help you better exercise AI's potential in your personal and professional endeavors. It teaches you how to unlock AI's full potential. We guide you through the art of AI communication and give you powerful tools to master these interactions with confidence and skill.

Throughout these pages, you'll find:

1. A foundation in the fundamental concepts of conversational AI
2. Practical techniques for crafting effective prompts and managing AI attention
3. Strategies for leveraging AI in various fields, from content creation to scientific research
4. Ethical considerations and best practices for responsible AI use
5. Insights into the future of human-AI collaboration and how to prepare for it

Target Audience

This book is written for a diverse audience, including:

- AI practitioners who want to upskill in AI discussion techniques
- Professionals seeking to integrate AI into their work processes
- Students and researchers in AI-related fields
- Business leaders and decision-makers exploring AI adoption
- Anyone curious about enhancing their interaction with AI systems

Whether you're a novice just beginning to explore the world of AI or an AI expert, this book will provide valuable insights and applicable techniques.

How to Use This Book Effectively

To get the most out of "AI Conversations That Work: An Ethical Professional's Guide to Directing Artificial Intelligence," we recommend the following approach:

Read sequentially: The chapters are designed to build upon each other, providing a logical progression from basic concepts to advanced applications.

Engage with the exercises: The book contains practical exercises and examples. Take the time to work through them—they are a pragmatic way to learn by doing.

Reflect and apply: After each chapter, take a moment to reflect on how the concepts and techniques may improve your productivity with AI. Consider keeping notes on your thoughts and ideas.

Experiment with actual AI systems: As you progress through the book, practice the techniques with AI models. This hands-on experience is invaluable for deepening your understanding.

Revisit and review: Consider revisiting earlier chapters as you encounter more advanced concepts. The interconnected nature of AI communication means that earlier principles often take on new significance as your understanding grows.

Engage with the community: Use the resources in the appendices to connect with other AI enthusiasts and professionals. Sharing experiences and insights can significantly enhance your learning journey.

In Closing, Let's Begin

Remember, the field of AI is rapidly evolving. While this book provides a solid foundation, staying curious and learning beyond these pages will be critical to your long-term success in AI communication.

As you embark on this journey, approach it with an open mind and a spirit of experimentation. The art of AI communication is as much about creativity and intuition as it is about technical knowledge. Embrace the challenges, celebrate the breakthroughs, and enjoy the process of mastering this transformative skill.

Welcome to the exciting world of AI communication. Let's begin!

1

OVERVIEW

1.1 Introduction: The New Era of AI Communication

AI needs no introduction, not anymore. By now, you've probably already chatted with at least one AI, like ChatGPT by OpenAI or Claude by Anthropic. Likely, readers have already engaged in multiple interactions with AI systems. Many may have even utilized AI to generate substantial project outputs, such as software or team documentation. One common issue people face with AI is that the AI will seem quick to forget things that you tell it – or even stuff that it tells you.

The two services I mentioned, ChatGPT and Claude, employ multiple Large Language Models (for example, Claude Sonnet 3.5). Large Language Models are commonly referred to as LLMs, and LLMs are the digital brains at the heart of the current generation and the most popular flavor of AI today. Some readers will already have developed their style to get what they want from these commonly used AIs if not alternatives such as Llama 3.2 by Meta. You can further hone your style of talking to AI and doing so will help get more of what you want more efficiently. I encourage you to share your acquired knowledge with the broader community through various mediums, such as publications, blog posts, or relevant online forums.

With the recent advances in the LLM space, sometimes called "chat AI," the stage is set for you to do great things with AI's immense intellectual and creative powers. Access to advanced AIs such as ChatGPT and Claude is widely available. You can often find an AI that is free or relatively low-cost to use.

The overarching theme of this book is developing practical and valuable outputs (read as beyond demo-quality outputs) through attention-directed discussion with AI.

I initially set out to generate the entire book from a very dense brain dump, which I fed into ChatGPT O1-preview in a single pass. When that didn't work, I fed the initial results from O1-preview into Claude Sonnet 3.5 several times in iterative rounds, attempting to improve it and make it more cohesive. Sadly, those attempts didn't work as well as I hoped they would.

The tech isn't quite "there yet" concerning generating such large and complex outputs in a one-shot (single message input) pass. Also, the tone and depth of what AI generates leave much to be desired. Therefore, this book required a significant investment of my time, as I had to alter sections repeatedly by hand to get the book published. Although AI made the whole process much more fun and far less intimidating, this book did not spring forth from the AI in full as published.

My initial attempt at writing this book using only clever prompting with a brain dump failed. The result failed to meet my needs because I had so much content of little cohesion that I had to start over from scratch. The high-quality, relatively casual tone, medium depth, and lock-step cohesiveness I desired were ultimately only available to me by jumping in and rewriting sections of the book, sometimes more than once. My first attempt yielded about 150 pages of a generic "book about AI." The generic content failed to faithfully reflect my core ideas. AI also broke from my desired tone and crept far beyond the scope I wanted for the book.

After deliberating on what went wrong the first time around, I realized that some but not all aspects of the poor quality of the first attempt could be mitigated by even more specific and consistent prompting. Not getting what you want from AI for lack of not being clear or consistent in your requests is to be expected with any generation of AI. This requirement will remain true beyond the present generation of AI.

What you're reading is a hand-crafted editing of my second attempt at generating the book via AI. This second attempt was much more deliberate and structured in how I used AI to build out the content. I expanded my original brain dump and made it more detailed before involving AI again. I was very explicit about where I wanted to take the book.

I generated the table of contents iteratively. The first paragraph of each chapter was generated and refined. The chapters were generated by applying the same approach to each chapter, and by that, I mean verbatim the same prompt except for changing the chapter number desired.

Once I had all the chapters in basic form from the AI, I polished each by hand. With the help of Grammarly (see grammarly.com) and Claude, I edited the book. Grammarly can apply rules-based and AI-based editing on a fine-grained scale, and it is a wonderful tool for tracking down references for what the AI draws upon.

So then, even employing many of the techniques presented in this book, and even more techniques not mentioned, today's AIs will often run far astray toward ideas out of scope such that their outputs frequently need manual changes to align them with your goals.

To avoid changing AI outputs as much as I chose to, I might have written a tailor-made agent-based application. Agent-based AI applications stitch several parameterized AI conversations together in useful sequences. With such an app, what I chose to do manually in the writing of this book could perhaps have been done far more automatically.

Creating such an advanced AI-driven application would have achieved far more repeatable results than the turn-based discussion approaches detailed in the book. However, the task of creating the application would have set back the publishing of this book by months if not years. A significant amount of scientifically driven prompt engineering would be necessary to automatically yield such results reliably.

Prompt engineering at the scale of generating publishable books in a single-shot requires more substantial upfront investment than I had to lend to the task. By the time I would have created and tuned the fully automated book generating tool, the industry would probably have already done the same, and my book would go to market too late for my needs. Sooner than later, my goal is to help you develop greater intuition and skills for generating useful AI outputs through firsthand discussions with AI instead of by way of obsessively constructed and code-programmed chains of prompts.

AI is powerful, but sometimes, it still requires a prohibitive amount of human effort to solve the problems you throw at it. For this reason, I want to help you become an expert at directing AI attention across the flow of your chats with it. By approaching your discussions with AI as an artisanal craft, you can reduce the risk of needing to repeatedly restart your AI-generation-assisted projects, and move onto efficiently polishing AI outputs you already generated.

If you direct AI to edit one of your projects in full (for example, a book, script, or CSV), AI may subtly corrupt your project's structure or data in parallel to enriching it. Always maintain skepticism in what you receive from AI so that you stay in the loop and ensure quality from the ground up.

All subsequent chapters of this book were generated and then tuned by hand. The remaining chapters of this book are my human editorialization of what AI generated from what I told it. The original brain dump included my thoughts and intentions for this book. In the end, I was able to weave in a far more nuanced level of my thoughts and ideas by editing the AI-generated outputs. This process of developing in a cyclical flow with AI involved is commonly called "Human in the Loop" AI-based development, and it's the best way I know to solve challenging problems and massive complex outputs using AI.

It is best to iterate and tune your AI inputs in the earliest stages of developing large AI-assisted projects. As you draw closer to your end goal, you should instead iterate on AI outputs. Building out your project this way reduces the risk of introducing catastrophic corruption by AI dropping already polished content by mistake or misjudgment.

In practice, applying this guidance is relatively straightforward. Suppose you paste a project planning brain dump into the AI chat box and click send. You may receive only a summary of your input from the AI, or the AI may tell you that it has read what you sent and then ask your request.

You would be better served if your first input were more direct and assertive. In this way, you can build up to a large base of generated content and carve out the desired end state from that rock.

For any interaction you conceive of with AI, go about it in a way that will allow you to do it several times in a loop. If you approach the building out of large projects in this way, the project will materialize iteratively, playing to your strengths and those of the AI.

In summary, and as always has been true about computers, with most, if not all, things computer-related, garbage input begets garbage output.

Now begins the book's human-prompted, AI-generated, and human/AI-edited portion. Thank you very much for reading this book.

1.2 A Flashlight in Darkness

Effectively interacting with AI systems can be conceptualized as managing an extensive knowledge repository. This repository encompasses human understanding, including all facts, concepts, and ideas developed throughout history. The ability to focus attention within this vast information landscape is analogous to exploring this library with a flashlight in the dark.

This section introduces techniques for optimizing the selection and presentation of knowledge when interacting with AI systems. The emphasis is not on the breadth of information available or the AI's processing power but on developing a nuanced understanding of which pieces of information are most relevant and valuable for the AI system at any given moment.

A critical aspect of effective AI communication lies in the human operator's deliberate selection and emphasis on particular knowledge. When presented with a query or task by an AI system, the process involves making rapid, strategic decisions about which facts to prioritize, which concepts to elucidate, and in what sequence to present this information. This approach differs significantly from a random or unfocused presentation of data.

By honing this skill you'll be able to:

- Pinpoint the most relevant and accurate information for any AI task (no more information overload!)
- Highlight unexpected connections that lead to creative and tailored outputs (goodbye, predictable responses!)
- Illuminate a perfect sequence of knowledge to solve complex problems efficiently (hello, eureka moments!)
- Keep your AI focused on pertinent information, avoiding the shadows of irrelevance (farewell, AI tangents!)

Throughout this book, we'll explore practical techniques for selecting the most impactful knowledge to illuminate, managing the information you provide to AI, leveraging different domains of your vast knowledge repository, and much more. Each chapter builds on this core concept, providing actionable strategies to enhance knowledge-illuminating skills.

Why is this so crucial?

As AI becomes more ubiquitous, your ability to quickly identify and highlight the most relevant knowledge will be as important as critical thinking is today. Whether you're a professional looking to enhance your productivity, a researcher aiming to break new ground, or someone interested in getting the most out of AI tools, this book will equip you with the insights and techniques you need.

By mastering the art of Knowledge Illumination (KI) for AI, you'll be able to:

- Skyrocket your productivity by instantly spotlighting the most relevant information for any task
- Unlock new realms of creativity by highlighting unusual connections between disparate pieces of knowledge
- Make more informed decisions by expertly illuminating the most crucial data points amidst a sea of information
- Stay at the forefront of innovation by always knowing which cutting-edge concepts to bring to light

So, are you ready to refine your knowledge-illuminating skills and learn how to shine a light on precisely the correct information at the right time? Let's dive in and start transforming the way you interact with AI. Trust me, once you master this, you'll see the power of your knowledge in a whole new light!

1.3 It's All in YOUR Head: The Core Concept

The core concept of this book is dead simple, yet it'll blow your mind: "It's all in YOUR head." This isn't just a catchy phrase; it's the fundamental principle that will revolutionize how you approach productivity with AI.

When interacting with an AI, you're not merely typing text into a box —you're orchestrating a sophisticated pattern-matching system. Think of it like directing a spotlight: You determine where the AI focuses its analytical capabilities. By visualizing the AI's attention as this moveable spotlight, you can strategically guide it to relevant information and concepts. Understanding what information is currently 'illuminated' in the AI's context window is crucial for productive collaboration.

Think of the AI's mind as a colossal library, with millions of books (data points) on countless shelves (neural networks). When interacting with the AI, you guide a librarian through this massive collection. Your prompts, questions, and inputs are like light from a flashlight, illuminating certain library sections while leaving others in darkness.

But here's the crucial part: the library layout, the way the books are organized, the paths between the shelves - all of this exists in your imagination. By developing a rich, detailed mental model of how the AI processes information, you gain the power to guide it more effectively.

Let's consider a concrete example:

Imagine you're using an AI to help you brainstorm ideas for a new eco-friendly product line. A novice might ask, "Give me ideas for eco-friendly products." But armed with your mental model of the AI's attention, you approach it differently:

> **You:** "Let's explore ideas for eco-friendly products. First, I want you to focus on current environmental challenges. Highlight three major issues affecting our planet."
>
> **AI:** [Provides information on three environmental issues]
>
> **You:** "Great. Now, keeping those challenges in mind, let's shift our attention to existing eco-friendly technologies. What cutting-edge sustainable materials or processes have emerged in the past year?"
>
> **AI:** [Offers information on recent eco-friendly innovations]
>
> **You:** "Excellent. Now, I want you to imagine a Venn diagram. In one circle, we have the environmental challenges. On the other, we have these new technologies. I want you to focus on the overlap - where these new technologies could directly address our environmental challenges. From this intersection, generate five product ideas."

In this interaction, you're not just asking for ideas, you're actively shaping the AI's attention, guiding it through a thought process, and directing its focus to specific areas of its knowledge. This is the power of having a rich mental model of the AI's cognition - it's all in your head. Maintaining a mental model of AI's attention throughout your AI conversations allows you to orchestrate your interactions with greater precision and creativity than leaving AI's focus to chance and its training.

1.4 The Art of Knowledge Illumination

As we progress through these chapters, we'll be developing what earlier we called "Knowledge Illumination" - the art of guiding these digital minds with finesse and precision. This isn't just about getting better outputs from AI; it's about fundamentally changing how you approach problem-solving and creativity in the age of artificial intelligence.

Remember, the AI models we're working with are not just glorified search engines. They're complex, dynamic systems capable of reasoning, creativity, and surprising insights. But they're also, in a sense, very literal. They are also non-deterministic in many cases, such that becoming an artisan in how you discuss things with AI is a justifiable goal. AIs must have the broader context and common-sense understanding that humans naturally bring to conversations.

This is where your mental model becomes crucial. By visualizing the AI's attention, you can guide it more effectively. You can help it focus on relevant information, make connections it might miss, and even encourage it to approach problems from novel angles.

Let's expand on our earlier example to illustrate this:

> **You:** "Based on our previous discussion about eco-friendly product ideas, let's focus on the concept of biodegradable packaging. I want you to imagine you're a materials scientist specializing in biodegradable polymers. From this perspective, what are three cutting-edge developments in this field that could revolutionize packaging?"
>
> **AI:** [Provides information on three biodegradable polymer developments]
>
> **You:** "Excellent. Now, let's shift perspectives. Imagine you're a marketing expert focused on consumer behavior around eco-friendly products. How might these new packaging technologies appeal to environmentally conscious consumers? What potential objections or concerns might they have?"
>
> **AI:** [Offers insights on consumer perspectives and potential concerns]
>
> **You:** "Great insights. Now, let's combine these perspectives. Suppose you were a product developer tasked with creating a new line of eco-friendly food containers using one of these biodegradable polymers. What would be your top three considerations in designing the product?"

Here again, you're not just asking for information or ideas, you're actively shaping the AI's perspective, guiding it to approach the problem from different angles and synthesizing these viewpoints. This is the essence of Knowledge Illumination—using your understanding of the AI's cognition to orchestrate a multifaceted exploration of ideas.

1.5 Conclusion: It's a Human-AI Collaboration

You'll develop an intuitive feel for this dance of mind as we progress through this book. You'll learn to sense when the AI is on the right track and how to correct its course when it strays. You'll discover how to use the AI's vast knowledge and processing power as an extension of your intellect, amplifying your problem-solving and creative capabilities.

Remember, the goal isn't to replace human thinking but to augment it. The most powerful insights often come from combining human intuition and AI processing. By mastering the art of Knowledge Illumination, you're not just learning to use a tool - you're expanding the boundaries of your cognitive capabilities.

In the following chapters, we'll delve deeper into the techniques and strategies that make this possible. We'll explore the nuts and bolts of AI cognition, develop practical skills for guiding AI attention, and discover how to apply these skills across various domains.

Let's continue exploring the fascinating world of human-AI collaboration.

2

DEMYSTIFYING AI-LANGUAGE MODELS

2.1 Introduction: The Magic Behind AI Conversations

Imagine conversing with a librarian that's read every book ever written. Now, picture this librarian not just recalling information but also understanding context, generating new ideas, and even cracking jokes. This is what an AI language model does. In this chapter, we will demystify these fascinating digital minds we've been learning to collaborate with.

2.2 What Are AI Language Models?

At their core, AI language models are sophisticated pattern recognition machines. They are trained on gigantic datasets, learning the intricate patterns and structures of language. However, unlike a simple pattern-matching system, these models have developed a nuanced understanding of language, allowing them to generate coherent and contextually appropriate text.

Let's break it down with a fun analogy:

Imagine playing a massive "word association" game with a very intelligent friend. You say a word, and they respond with another related word. Now, expand this game to include phrases, sentences, and paragraphs. Your friend doesn't just react with memorized phrases but understands the context and can create new, relevant responses. That's what an AI language model does!

It's really a bit like how Bumblebee, a character in the Transformers fictional universe, strings together bits of radio content in order to speak without having a more traditional voice.

As we discuss language models, remember that their attention mechanisms (covered in detail in Chapter 3.1) are crucial to their performance. Understanding these fundamentals will be especially relevant when we explore advanced prompting techniques in Chapter 4.2.

Chapter 5.2 contains hands-on exercises that demonstrate how different model architectures affect brainstorming outcomes, allowing you to apply these concepts practically.

2.3 The Role of Attention

Attention mechanisms are fundamental to how modern AI processes information. While we'll explore the technical details in Chapter 3.1, here's what you need to know:

Modern AI uses attention to:

- Focus on relevant parts of the input, like a human focusing on critical points in a conversation
- Make connections between different pieces of information, similar to how we connect related ideas
- Process context effectively by keeping track of important information throughout a conversation

For instance, when you ask about "it" in a sentence, attention helps the AI understand what "it" refers to by focusing on relevant earlier context. This ability to maintain and connect context is crucial for meaningful conversations with AI.

> ***Tip:*** *For a deeper technical dive into attention mechanisms, see Chapter 3.1.*

2.4 The Evolution of Language Models: From Simple to Spectacular

Let's take a quick journey through the evolution of language models.

It's like watching a child grow into a genius!

Early Days: The first language models were like simple word prediction tools. Think of the autocomplete on your phone, but much more essential.

Growing Smarts: As models grew, they better understood longer text sequences and more complex patterns.

The Transformer Revolution: In 2017, a new type of model called a Transformer changed the game. It was much better at understanding context over longer passages.

Modern Marvels: Today's models, like GPT-3 and Claude, can understand and generate human-like text on almost any topic. They can write stories, answer questions, and even help with coding!

2.5 Let's Play: Experimenting with Language Models

Now that we understand the basics, let's have some fun experimenting with language models.

Here are a few activities you can try:

Completion Challenge: Start a sentence and see how different AI models complete it. For example: "In a world where gravity suddenly reversed..."

Style Shift: Ask an AI to rewrite a paragraph in different styles—like Shakespeare, a news reporter, or an excited athlete.

Language Leap: If you know multiple languages, try having a bilingual conversation with an AI and see how well it switches between them.

Idea-Expansion: Give the AI a simple concept and ask it to expand into a short story or explanation.

Remember, the goal is to understand the AI's capabilities and limitations. You might be surprised at how human-like (and sometimes amusingly non-human) the responses can be!

2.6 The Superpowers and Kryptonite of Language Models

Like any superhero, AI language models have their strengths and weaknesses. Let's explore them:

Superpowers:

Vast Knowledge: They can access and synthesize information on almost any topic.

Creativity: They can generate original content, from poems to code.

Language-Understanding: They can understand and generate text in multiple languages.

Contextual-Awareness: They can maintain context over long conversations.

Kryptonite (Limitations):

Lack-of-True-Understanding: Despite their abilities, they don't truly "understand" as humans do.

Potential-for-Biases: They can reflect biases in their training data.

Hallucinations: Sometimes, they can generate plausible-sounding but incorrect information.

No-Real-time-Updates: Their knowledge is often limited to their training data cutoff date which is commonly several months prior to your discussion with the AI.

2.7 Ethical-Considerations: The Responsibility of Power

With great power comes great responsibility, and AI language models are no exception.

HERE ARE SOME KEY ETHICAL CONSIDERATIONS:

Privacy: How do we ensure that personal data used in training or interactions remains protected?

Bias: How can we prevent and mitigate biases in AI language models?

Misinformation: How do we prevent the spread of AI-generated false information?

Transparency: How can we make AI decision-making processes more transparent and explainable?

These are complex questions without easy answers, but they're crucial to consider as we continue to develop and use AI language models.

2.8 The Future is Conversational: What's Next for Language Models?

As we look to the future, the possibilities are exciting:

Multimodal-Models: Future AI might seamlessly integrate text, images, and even video understanding.

Enhanced-Personalization: AI could adapt more precisely to individual users' communication styles.

Improved-Reasoning: We might see AI that can perform more complex logical reasoning tasks.

Ethical-AI: Future models could have more robust built-in ethical guidelines.

The journey of AI language models is just beginning, and the future promises even more fascinating developments!

2.9 Conclusion: Your AI Collaboration Journey Begins Here

Understanding AI language models is the first step in mastering the art of AI collaboration. As we move forward in this book, we'll build on this foundation, exploring practical techniques for harnessing the power of this revolutionary technology.

Remember, AI language models are incredibly sophisticated tools, but tools nonetheless. Human creativity, critical thinking, and ethical judgment will unlock their potential. So, armed with this new understanding, let's continue our journey into the exciting world of AI collaboration!

CHAPTER 2 SUMMARY: DEMYSTIFYING AI-LANGUAGE MODELS

Key Concepts:

1. The fundamental nature and workings of AI language models
2. Evolution from simple language models to modern AI systems
3. Key capabilities and limitations of language models
4. Ethical considerations in using AI language models
5. The future trajectory of conversational AI

Reflection Questions:

1. How might understanding the inner workings of language models influence your approach to AI collaboration?
2. What challenges do the current limitations of language models present for your specific use cases?
3. How can you balance leveraging AI capabilities while remaining mindful of ethical considerations?
4. What implications does the evolving nature of language models have for long-term project planning?

Action Items:

1. Experiment with different language models to understand their unique characteristics
2. Create a list of potential use cases specific to your field, considering both capabilities and limitations
3. Develop a practical framework for ethical AI usage in your projects
4. Set up a system to stay informed about new developments in language model technology

Further Exploration:

- Research specific language models being used in your industry
- Investigate technical documentation for models you frequently use
- Study recent papers on language model architectures and capabilities
- Explore emerging ethical guidelines for AI language model usage

3

INSIDE THE AI MIND: ATTENTION AND CONTEXT

3.1 Introduction: The Attention Mechanism Explained

Let's take a friendly tour of how AI models work, focusing on their most crucial component: the attention mechanism. This innovation enables AI to process and generate human-like text with remarkable understanding.

The attention concepts we're exploring form the foundation for many practical techniques we'll cover in Chapter 4. In particular, the "push-pull" dynamics discussed in Section 3.5 directly inform the prompt engineering strategies in Section 4.2 and the creativity techniques in Chapter 5.3.

> ***Tip:*** *For readers particularly interested in attention mechanisms, Chapter 7.2 explores how these concepts apply to Claude's Artifacts feature, while Chapter 12.2 examines their role in next-generation multimodal systems.*

Basic Operations

First, let's understand the fundamental process:

Input Processing: When you type a message to an AI, it first breaks down your text into smaller pieces called tokens. These could be words, parts of words, or even punctuation marks.

Context Understanding: The AI then looks at these tokens and tries to understand the context. It's like reading a book and tracking what's happening in the story.

Probability Magic: The AI analyzes patterns in its training data to calculate probability distributions for probable subsequent tokens (words or word parts). This process involves complex mathematical operations across multiple neural network layers, considering local context (nearby words) and broader context (overall meaning). While similar to predicting the next word in a sentence, it operates on multiple levels simultaneously - vocabulary, grammar, semantics, and domain knowledge.

Generation: The AI responds based on these probabilities. It creates new text that makes the most sense given the context.

Fun Fact: *This process happens lightning-fast, often in mere milliseconds!*

Understanding Attention

The attention mechanism is what makes this process truly powerful. Think of it like a skilled librarian who can instantly identify and rank relevant information from a colossal databank. This librarian doesn't just recall information sequentially but can dynamically weigh the importance of different pieces of information based on the current context.

In technical terms, the attention mechanism allows an AI model to create direct connections between different positions in a sequence of input data, regardless of distance. This ability enables the model to consider long-range dependencies and contextual nuances that might be lost in more basic, sequential processing methods.

The Interplay of Queries, Keys, and Values

In our grand library analogy, we can think of the attention process in these terms:

> **Queries:** Our librarian is trying to address specific questions or topics. Each query is a search for relevant information.
>
> **Keys:** These labels or indexes on our library's books and passages. They help identify what information is available and where it can be found.
>
> **Values:** These represent the actual content of our books and passages - the substantive information that will be used to formulate an answer.

The AI processes information and performs a sophisticated search and retrieval operation. The query (our specific information need) is compared against all the library Keys (labels of available information). When a relevant Key is found, the corresponding Value (the actual content) is brought into focus.

Self-Attention: The Library's Cross-Referencing System

One potent form of attention is self-attention, where the model attends to different parts of its internal representation. In our library analogy, this is akin to an advanced cross-referencing system.

Imagine our librarian searching for books related to the initial query and constantly checking how each piece of information relates to and informs other pieces. This self-referential process allows the librarian (our AI model) to build a rigorous, contextualized understanding of the topic.

This cross-referencing enables AI to capture long-range dependencies in language, understanding, for instance, that a pronoun like "it" in one sentence refers to a subject mentioned several sentences prior. In library terms, it's like understanding that a brief note in one book is a crucial reference to a concept detailed in another book entirely.

This self-attention mechanism gives modern AI its remarkable ability to navigate context and generate nuanced language, much like how a skilled librarian can provide insights by connecting information from various sources across the library.

3.2 Understanding the Context Window

Imagine peering through a magical window that lets you see a certain number of words simultaneously. As you move this window across a text, you can only consider the words visible within its frame at any moment. This is the context window in AI language models.

The context window is the range of text that the model can "see" and consider when generating or processing language. It's like the AI's short-term memory, containing the task's most recent and relevant information. The context window size depends on the choice of AI model, with more advanced models generally having more oversized windows.

The Ebb and Flow of Information

Older information is inevitably pushed out as new information flows into the context window. This creates a dynamic environment where the relevance and importance of information are constantly in flux. It's akin to a conversation where recent statements are fresh in mind, but earlier points may fade unless deliberately brought back into focus.

Understanding how to deliberately bring information into the AI's attention span is crucial for effective communication. Remaining mindful of the state of the AI's focus during each discussion and with each response helps us structure our prompts and questions to keep the most critical information within the AI's "view." This, in turn, helps us improve the productivity of our discussions with AI.

Techniques for Managing Context

Chunking: Breaking long inputs into smaller, manageable pieces that fit within the context window.

Summarization: Periodically ask the AI to summarize critical points, returning essential information to focus.

Explicit Referencing: Directly referring to earlier parts of the conversation to promote them in the context window before they pass out of it.

3.3 Memory Stores and Retrieval-Augmented Generation (RAG)

While the context window serves as the AI's short-term memory, more persistent forms of "memory" have been developed to enhance AI's capabilities. Enter Retrieval-Augmented Generation, or RAG—a technique akin to giving our AI librarian access to a vast external archive of information.

RAG enhances large language models by pairing their text generation prowess with dynamic access to external data sources. RAG provides the large language model with a searchable memory bank to reference before crafting its responses.

It's like having a brilliant conversationalist who, mid-dialogue, can instantly consult a wiki to enrich the discussion with pertinent facts and details.

The Power of External Knowledge

RAG allows AI models to access and utilize information beyond what's contained in their trained parameters. This approach is handy for tasks that require up-to-date or specialized knowledge, helping to mitigate the problem of AI "hallucination" or generating incorrect information.

Imagine you're discussing recent scientific discoveries with an AI. With RAG, the AI can pull in the latest research findings from a datastore that indexes them, better ensuring the conversation is grounded in current knowledge rather than potentially outdated training data.

Implementing RAG in Conversations:

To leverage RAG effectively, where it is enabled by your chat AI, consider these strategies:

Explicit Knowledge Requests: Asking the AI to check its external knowledge base for specific information.

Fact-Checking Prompts: Encouraging the AI to verify claims against its external sources.

Contextual Enrichment: Request additional background information on topics to deepen the conversation.

> ***Note:*** *Not all AIs will have external facilities available to them.*

3.4 Tool Usage: Extending AI Capabilities

In our quest to create more capable AI assistants, we've equipped them with various tools, much like a Swiss Army knife of digital capabilities. These tools extend the AI's reach beyond mere text generation, allowing it to perform multiple tasks that would otherwise be beyond its grasp.

Imagine an AI that can discuss the weather and check real-time weather data for any location. Or an AI that can talk about mathematics and solve complex equations on the fly. This is the power of tool usage in AI.

The AI's Toolbox

AI tools fall into distinct categories with specific capabilities: computational tools (for mathematical operations and data analysis), information retrieval tools (for accessing current data through Application Programming Interfaces, known commonly as APIs), functional tools (for specific tasks like code execution or format conversion), and integration tools (for connecting with external software systems). Each tool type requires specific prompting strategies and an understanding of its limitations.

Some standard tools include:

- Web search engines
- Mathematical computation engines
- Data visualization tools
- Code execution environments
- Calendar and scheduling systems

Orchestrating Tool Use

The true art lies in the AI's ability to orchestrate these tools, knowing when and how to deploy them. It's like a conductor leading an orchestra, bringing in different instruments (tools) at the right moment to create a harmonious performance (its response).

3.5 Shaping AI Attention: The Art of Pushing and Pulling

Now that we've discussed the inner workings of AI attention and context, let's explore how users can actively shape AI's focus. This is where the concepts of "pushing" and "pulling" come into play.

Pushing: Directing the AI's Focus

Pushing involves actively directing the AI's attention to specific information or concepts. It's like using a highlighter to emphasize key points in a text.

Here are some techniques for effective pushing:

Explicit Statements: Clearly state the most critical points you want the AI to consider. For example: "Pay special attention to the following three points..."

Repetition: Repeat crucial information at strategic points in your input. This reinforces its importance in the AI's attention mechanism.

Formatting: Use capitalization, bold text symbols such as those used in the markdown text format, or other formatting to emphasize key points. For instance: "The MAIN GOAL of this project is..." (where the two words "MAIN GOAL" are in all capital letters)

Structured Input: Organize your input with clear headers, bullet points, or numbered lists. This helps the AI parse and prioritize information.

Emotional Language: Use emotionally charged or vivid language for crucial points. The AI often pays more attention to emotionally salient information.

Emojis: In a message where you give constructive feedback to an AI, using a frown emoji with a quantitative multiplier like frown times 1 (where the most recent output is not quite what you want) versus frown times 1000 (when the most recent output is very far off-track from your request) is a surprisingly effective way to escape cycles in which the AI repeatedly provides similarly incorrect responses.

Pulling: Guiding the AI to Retrieve information

Pulling involves encouraging the AI to retrieve or focus on specific information from its knowledge base or earlier in the conversation. It's like asking our librarian to fetch a particular book from the shelves. Here are techniques for effective pulling:

Direct Questions: Ask specific questions about the information you want the AI to focus on. For example: "What did we discuss about the attention mechanism earlier?"

Analogies and References: Use analogies or references to concepts you've discussed before. This encourages the AI to draw connections and pull relevant information into focus.

Contextual Prompts: Provide context that hints at the type of information you want the AI to retrieve. For instance: "Considering what we know about neural networks..."

Hypothetical Scenarios: Present hypothetical situations that require the AI to apply specific knowledge, pulling it into the forefront of its processing.

Contrast and Comparison: Ask the AI to compare or contrast ideas, naturally pulling relevant information into focus.

The Push-Pull Dance

Mastering the art of pushing and pulling allows you to engage in a sophisticated dance with the AI, guiding its attention and shaping the conversation. It's like leading in a dance–you provide direction and cues, and the AI responds, creating a fluid and productive interaction.

For example, you might push important context at the beginning of a conversation:

> "We're discussing renewable energy technologies. The THREE KEY FACTORS to consider are efficiency, cost, and environmental impact." (In this message to the AI we typed the words, "THREE KEY FACTORS" in all caps.)

Then, later in the conversation, you can pull this information back into focus:

> "Considering the three factors we mentioned initially, how would you evaluate solar power versus wind power?"

This push-pull dynamic lets you control the conversation's direction while leveraging the AI's vast knowledge and processing capabilities.

ONE THING that is absolutely key to know about the AI you choose is: how far back in a discussion will your AI recall. While you can fairly easily find the context window size, which is most commonly measured in tokens, for most AI models in their documentation, developing effective intuition about how far back in a discussion your AI will recall is best done through practice and experimentation with both fresh and long conversations.

3.6 Practical Exercises in Shaping AI Attention

To help solidify these concepts, let's engage in a few practical exercises.

TRY THESE WITH YOUR FAVORITE AI ASSISTANT:

The Highlight Reel: Write a paragraph about a topic of your choice, then ask the AI to summarize it. Next, rewrite the paragraph with key points emphasized (using caps, bold, etc.), and ask for another summary. Compare the results to see how emphasizing (pushing) certain information affects the AI's focus.

The Memory Game: Have a multi-turn conversation with the AI about a complex topic. Periodically ask it to summarize critical points discussed so far. Notice how this pulling technique helps maintain important information in the context window.

The Tool Tango: Present the AI with a problem requiring multiple tools (for example, a math problem requiring current data). Practice guiding the AI to use different tools in a linear sequence, observing how it integrates the results into a cohesive response.

The Analogy Bridge: Introduce a concept early in the conversation using a vivid analogy. Later, reference the analogy to pull that concept back into focus. Observe how the AI uses the analogy to reconnect with the earlier information.

The Context Window Shuffle: In a long conversation, deliberately bring up information from much earlier in the discussion. Notice how the AI handles this - does it still have that information in its context, or must it be reminded? You can always ask the AI to tell you what it recalls from earlier.

3.7 Limitations of Current Attention Mechanisms

While general AI limitations are covered in Chapter 2.6, attention mechanisms specifically face these unique challenges:

Context Window Size:

- Attention becomes computationally expensive with longer inputs
- Current models can only "pay attention" to a limited amount of context

Attention Dilution:

- With very long inputs, attention can become spread too thin
- Important information might get lost among less relevant details

Cross-Document Limitations:

- Difficulty maintaining attention across multiple separate documents
- Challenges in synthesizing information from multiple sources

Computational Complexity:

- The longer your input, the exponentially harder it becomes for AI to process it effectively
- This creates practical limits on what you can achieve in each discussion with AI

THESE ATTENTION-SPECIFIC LIMITATIONS inform how we structure our interactions with AI systems.

3.8 Real-World Applications of Attention in AI Systems

Attention mechanisms have found applications across a wide range of AI tasks:

Machine Translation: Attention allows for better alignment between source and target languages, improving translation quality.

Text Summarization: Attention-based models can generate more accurate and relevant summaries by focusing on critical parts of a document.

Image Captioning: Attention helps models focus on relevant parts of an image when generating descriptive captions.

Speech Recognition: Attention mechanisms improve the accuracy of transcribing speech to text by allowing the model to focus on relevant audio segments.

Sentiment Analysis: Attention helps identify which words or phrases are most important for determining the sentiment of a piece of text.

Question Answering: Within QA systems, the attention mechanism is a precision tool that helps the model identify and concentrate on the most pertinent context segments when formulating accurate answers.

Code Generation: Attention mechanisms help AI understand the structure and context of code, enabling more accurate code completion and generation.

These applications demonstrate the versatility and power of attention mechanisms in enhancing AI's ability to process and generate information across diverse domains.

3.9 Case Studies: Attention Mechanisms in Popular AI Models

Let's examine how attention mechanisms are implemented in three popular AI models, along with examples of their applications:

BERT (Bidirectional Encoder Representations from Transformers)

BERT, developed by Google, employs a bidirectional attention mechanism, allowing it to simultaneously consider context from both the left and right sides of a sentence.

Application: Language Understanding in Search Engines

BERT has been integrated into search algorithms to enhance the understanding of user queries. For instance, in complex queries, BERT helps identify the context, improving the accuracy of search results related to specific user needs, such as visa requirements for travelers.

Key Features:

- Bidirectional attention
- Pre-training on masked language modeling and next-sentence prediction
- Fine-tuning for specific tasks

GPT (Generative Pre-trained Transformer)

The GPT series, developed by OpenAI, uses a unidirectional (left-to-right) attention mechanism, which allows it to generate coherent text by focusing on previous words.

Application: AI-Assisted Writing

GPT-3 has been widely used in content generation, where it assists users in writing articles, stories, and reports. For example, when prompted with a topic, GPT-3 can generate relevant text that discusses various aspects of that topic coherently.

Key Features:

- Unidirectional attention (autoregressive)
- Massive scale, allowing for nuanced responses
- Few-shot learning capabilities, enabling effective performance with minimal examples

T5 (Text-to-Text Transfer Transformer)

T5, developed by Google, frames all NLP tasks as text-to-text problems, utilizing attention mechanisms for both input and output.

Application: Multilingual Tasks

T5 is effectively used for tasks such as translation and summarization. It can process input text in one language and produce concise summaries or translations in another, showcasing its versatility across different NLP applications.

Key Features:

- Unified text-to-text framework for multiple NLP tasks
- Pre-training on a variety of tasks
- Attention mechanisms applied during both encoding and decoding stages

These examples illustrate how attention mechanisms, implemented differently, enable AI models to tackle a wide range of language understanding and generation tasks effectively.

3.10 The Future of Attention Mechanisms

Key developments shaping the future of attention mechanisms in AI:

Efficient Attention: Development of computationally efficient mechanisms to handle longer sequences and larger datasets.

Multimodal Attention: Integration of attention across different modalities (text, images, audio, video).

Hierarchical Attention: Implementation of attention at multiple levels of abstraction to improve reasoning capabilities.

Dynamic Attention: Adaptation of attention mechanisms based on specific tasks or inputs.

Explainable Attention: Development of attention mechanisms that clearly explain their focus and reasoning.

3.11 Integrating Attention in AI Interactions

Strategies to leverage attention mechanisms in AI interactions:

Structured Prompts: Use clear headings, bullet points, or numbered lists to guide the AI's focus.

Attention Priming: Explicitly state key concepts or context before asking complex questions.

Attention Checkpoints: Periodically ask the AI to summarize its understanding or focus on specific elements.

Cross-Reference Prompting: Encourage connections between different parts of the conversation.

Attention Debugging: Rephrase prompts to guide AI focus more effectively if responses are unexpected.

Multi-Turn Refinement: Use a series of prompts to refine the AI's attention and output progressively.

Attention Visualization: Ask the AI to explain which parts of the input it found most relevant.

3.12 Ethical Considerations in AI Attention

Important ethical aspects to consider:

Bias Amplification: Be aware of potential bias amplification and critically evaluate AI outputs.

Privacy Concerns: Consider privacy implications, especially when dealing with personal data.

Transparency: Ensure transparency in AI decision-making processes.

Overreliance: Avoid overestimating AI capabilities or relying on them without human oversight.

Dual Use: Be mindful of potential misuse in creating misinformation or deepfakes.

Cognitive Impact: Consider how interactions with AI systems might affect human cognitive patterns.

3.13 Conclusion: Harnessing AI Attention

As we conclude our exploration of AI attention mechanisms, let's recap the key insights:

Attention Fundamentals: Understanding how AI models focus on and process information is crucial for effective interaction.

Context Management: Recognize the limitations of the AI's context window and use techniques to maintain relevant information.

Strategic Prompting: Craft your inputs to effectively guide the AI's attention using structured prompts and explicit referencing.

Iterative Interaction: Engage in multi-turn conversations, refining and redirecting the AI's focus as needed.

Ethical Awareness: Remember potential biases and ethical implications when leveraging AI attention capabilities.

Continuous Learning: As AI technology evolves, stay informed about new developments in attention mechanisms and adjust your interaction strategies accordingly.

By applying these concepts, you can achieve better results and boost the efficiency and effectiveness of your engagements with AI systems. This approach allows you to harness their attention capabilities, leading to more precise, relevant, and insightful results in various applications.

CHAPTER 3 SUMMARY: INSIDE THE AI MIND: ATTENTION AND CONTEXT

Key Concepts:

1. Attention Mechanism: How AI models focus on specific parts of input data
2. Context Window: The range of text an AI model can consider when processing language
3. Push-Pull Dynamics: Techniques for guiding AI focus during conversations

Reflection Questions:

1. How might you apply the concept of "pushing" important information in your next AI interaction?
2. What challenges do you foresee in managing the context window during a long, complex conversation with an AI?
3. How does understanding the attention mechanism impact your approach to crafting effective prompts?

Action Items:

1. Practice crafting a prompt that explicitly directs the AI's attention to specific information
2. Experiment with the "pull" technique by referencing earlier parts of a conversation in follow-up queries
3. Design a multi-turn interaction that maintains important context throughout the conversation

Further Exploration:

- Research recent advancements in attention mechanisms in large language models
- Explore tools or techniques for visualizing AI attention in natural language processing
- Consider the ethical implications of directing AI attention in sensitive applications (e.g., healthcare, finance)

4

CONVERSING WITH AI: TECHNIQUES AND STRATEGIES

4.1 Introduction: The Art of AI Conversation

Welcome to the exciting world of AI conversation! This chapter explores the techniques and strategies to transform your AI interactions from simple Q&A sessions into dynamic, productive collaborations. Think of it as learning to dance with a partner who has access to the world's knowledge but needs your lead to create something beautiful.

Remember, AI isn't just a fancy search engine or a robotic assistant. It's a complex, dynamic system capable of nuanced understanding and creative output. As previously discussed, its power is unlocked through your ability to guide and shape the conversation. You're the conductor, and the AI is your orchestra!

4.2 The Art of Effective Prompting

While we explored the technical aspects of attention mechanisms in Chapter 3, we'll focus on their practical application in conversation.

Crafting a perfect prompt is like setting the stage for an outstanding performance. It's all about giving the AI the relevant context and direction to work magic.

These prompting techniques build upon our understanding of attention mechanisms (see Chapter 3.1).

The techniques mentioned in this section directly support the brainstorming methods we'll explore in Chapter 5.

For specific examples of how such techniques apply to different fields, see the case studies in Chapter 8, particularly Sections 8.2 (Content Creation) and 8.5 (Research and Development).

> ***Tip:*** *Remember the ethical principles discussed in Chapter 11.2 when crafting prompts. This becomes especially important when working with sensitive data, as explored in Chapter 9.6.*

Let's explore some techniques to make your prompts sing:

Be Specific and Clear

Imagine you're explaining a task to a brilliant but literal-minded coworker. That's your AI! Instead of asking, "What's the weather like?", try "What's the current temperature and forecast for New York City today?". This level of specificity helps the AI focus its vast knowledge on precisely what you need.

Use Context Wisely

Remember the context window we talked about in Chapter 3? Use it to your advantage! Set the stage for your AI like you're introducing a character in a play. For instance:

> "You are an expert in 19th-century literature. Analyze the themes in Jane Austen's 'Pride and Prejudice' from the perspective of a modern feminist critic."

This gives the AI a role and a specific lens to view the task.

Break Complex Tasks into Steps

Think of this as creating a recipe for your AI to follow. For complex queries, break them down into a series of smaller, manageable steps. For example:

1. Summarize the plot of 'Pride and Prejudice'.
2. Identify the main themes in the novel.
3. Analyze how these themes are developed throughout the story.
4. Discuss the relevance of these themes in modern society.

This approach helps you organize your thoughts and guides the AI through a logical progression.

Leverage the Power of Examples

Sometimes, showing is better than telling. Provide examples of the kind of response you're looking for. It's like giving your AI dance partner a quick demonstration of the steps. For instance:

> "Generate three marketing slogans for a new eco-friendly water bottle. Here's an example of the tone I'm looking for: 'Quench your thirst, not the planet's future.'"

Visualize the AI's Attention

Imagine you're directing the AI's spotlight of attention. Where do you want it to shine? If you're discussing a historical event, you might say:

> "Let's focus on the economic factors leading to the French Revolution. Imagine you're an economist from that era, analyzing the situation."

This prompt specifies the topic and guides the AI's 'attention' to a specific perspective and historical era.

Manage Multiple Copies of Important Information

In a long conversation, crucial information might get pushed out of the AI's context window. To mitigate this, consider adding multiple copies of vital information, especially with an explicit statement. It's like leaving breadcrumbs for the AI to follow. For example:

> "Important context (Copy 1): We are discussing the economic impact of the Industrial Revolution. [Your specific question here] Important context (Copy 2): Remember, our focus is on the economic impact of the Industrial Revolution."

This technique can help prevent "amnesia" in extended conversations, ensuring that crucial context remains in the AI's active memory.

4.3 Guiding AI Responses: The Dance of Collaboration

Now that we've set the stage with effective prompting let's explore how to guide the AI's responses. This is where the real magic happens – a dance of ideas between you and your AI partner.

Use Follow-up Questions

Feel free to ask for clarification or to dig deeper. It's like leading your dance partner into a new, more complex move. For example:

> **You:** "Explain the concept of quantum entanglement."
>
> **AI:** [Explains]
>
> **You:** "Fascinating! Now, how does this relate to quantum computing?"

Request Alternative Perspectives

Encourage your AI partner to look at things from different angles. It's like asking your dance partner to try a new style. For instance:

> "Now, explain quantum entanglement from the perspective of someone skeptical of quantum mechanics."

Encourage Creativity

While AIs excel at providing factual information, they can also be remarkably creative when prompted correctly. It's like encouraging your dance partner to improvise. Try something like:

> "Give me five unconventional uses for a paperclip in a post-apocalyptic world."

Use Hypothetical Scenarios

Hypothetical scenarios can be powerful tools for exploring ideas and concepts. It's like creating a new dance floor for you and your AI partner. For example:

> "Imagine a world where gravity suddenly became half as strong. How would this affect daily life and technology?"

Managing the Context Window

As you guide the conversation, be mindful of the AI's context window. Remember, it's like a sliding window of attention that moves as the conversation progresses. If you want to refer back to something mentioned earlier, reintroduce it. For example:

> "Earlier, we discussed the economic factors leading to the French Revolution. Considering those, how might a similar economic situation play out in a modern, globalized economy?"

This technique helps ensure that critical information stays within the AI's 'view' as the conversation evolves.

Balancing Specificity and Openendedness

While specificity is crucial for getting precise answers, sometimes you want to encourage more expansive thinking. Balance your approach by mixing specific questions with open-ended ones. For instance:

> **Specific:** "What were the three main causes of the French Revolution?"
>
> **Open-ended:** "How might the course of European history have changed if the French Revolution had never occurred?"

This balance lets you extract the AI's concrete information and imaginative insights.

4.4 Advanced Techniques: Leveling Up Your AI Collaboration

Let's explore some advanced techniques that will make your interactions even more powerful and productive.

Comparative Analysis Prompts

Challenge the AI to examine multiple angles of related concepts side by side. By asking for detailed comparisons rather than single-topic explanations, you'll uncover more profound insights into how different elements connect, overlap, and diverge from each other. This comparative approach helps build a more nuanced understanding of complex topics. It's like asking your dance partner to demonstrate the difference between a waltz and a tango. For example:

> "Compare and contrast the economic policies of Adam Smith and John Maynard Keynes. How might each approach address modern economic challenges?"

Socratic Questioning

Channel your inner Socrates! Use a series of probing questions to encourage deeper thinking and analysis. This technique can help uncover assumptions and explore implications. It's like leading your dance partner through an intricate series of steps, each building on the last. For instance:

> "What are the potential long-term impacts of renewable energy transition on global economies? What assumptions are we making about clean technology advancement? How might these transitions affect different regions and income levels?"

Constraint-Based Prompting

Introduce specific constraints or limitations to encourage creative problem-solving. This can lead to novel solutions and ideas. It's like challenging your dance partner to create a new routine using only three specific moves. For example:

> "Describe a sustainable city of the future, but with the constraint that no new technology beyond what we have today can be used. How would this city function, and what would daily life look like for its inhabitants?"

Chain-of-Thought Prompting

This technique involves asking the AI to break down its reasoning process step-by-step. It's beneficial for complex problem-solving tasks. It's like asking your dance partner to explain each step of a complicated routine. For instance:

> "Let's solve this calculus problem step-by-step. Start by identifying the type of problem, then outline the steps we'll need to take to solve it, and finally, work through each step, explaining your reasoning as you go."

Metacognitive Prompts

These prompts encourage the AI to 'think about its thinking,' which can lead to more nuanced and self-aware responses. It's like asking your dance partner to reflect on their performance mid-routine. For example:

> "Before you answer my next question, I'd like you to explain how you plan to approach it. What information will you need to consider? What potential biases might influence your response?"

4.5 Overcoming Misunderstandings: When the Dance Steps Don't Align

Even with the most carefully crafted prompts, misunderstandings can occur. Here's how to navigate these situations:

Identify the Misunderstanding

The first step is recognizing that a misunderstanding has occurred. The misunderstanding might be evident in off-topic responses and contradictory or potentially useless guidance. It's like noticing when your dance partner suddenly does the cha-cha while you're trying to waltz.

Rephrase and Clarify

If you suspect a misunderstanding, try rephrasing your question or statement. Use different words or structures to convey the same idea. As you describe your goals differently, you bring more of the neural network behind the AI into play.

For instance, if "What are the implications of quantum entanglement?" doesn't yield the response you're looking for, try "How does quantum entanglement affect our understanding of physics?"

Provide Additional Context

Sometimes, misunderstandings occur due to a need for more context. If this happens, take a step back and provide more background information. It's like reminding your dance partner of the choreography before focusing on a specific move. For example:

> "Let's back up a bit. We're discussing quantum entanglement in the context of secure communication systems. With that in mind, how does quantum entanglement contribute to quantum cryptography?"

Use the "Steelman" Technique

Instead of pointing out where the AI went wrong (a "strawman" argument), try the "steelman" approach. Restate the AI's response in the strongest possible terms, then clarify where your questions or needs differ. This can help reset the conversation constructively. It's like acknowledging your dance partner's interpretation of a move before suggesting a different approach.

Visualize and Realign the AI's Focus

When a misunderstanding occurs, it can be helpful to visualize where the AI's attention might have strayed and guide it back. You might say something like:

> "We may have drifted from our main topic. Let's refocus on the French Revolution's economic factors, particularly taxation's role. Can you summarize the key points discussed about taxation in pre-revolutionary France?"

This helps realign the AI's 'attention' with your intended topic, like gently guiding your dance partner toward the correct rhythm.

4.6 Leveraging Different Knowledge Sources: Expanding Your AI's Repertoire

As discussed in Chapter 3, AI can access different types of knowledge. Understanding how to tap into each type can significantly enhance your interactions. It's like knowing when to lead your dance partner into a ballroom waltz versus a street dance battle.

Intrinsic Knowledge

This is the knowledge baked into the AI during its training. It's broad but may only sometimes be up-to-date. To leverage this, you might say:

> "Based on your training data, what were the most influential social movements of the Renaissance period?"

External Knowledge (RAG)

You can guide the AI to use its Retrieval-Augmented Generation (RAG) capabilities for more current or specialized information. It's like asking your dance partner to incorporate a newly learned move. For instance:

> "Using your most up-to-date information sources, what are the latest developments in renewable energy technology?"

Tool-Based Knowledge

Remember that many AI systems can use external tools. You can explicitly request the use of these tools. It's like asking your dance partner to use props in the routine. For example:

> "Please use a calculation tool to determine the compound interest on a $10,000 investment at 5% annual interest over 20 years."

4.7 AI Troubleshooting Action Matrices

The following matrices provide immediate, actionable responses to everyday situations you'll encounter while working with AI.

When using these matrices:

1. Identify the closest matching situation
2. If the 'Do This' for your issue contains a double-quoted message, try applying the suggested solution precisely as written.
3. If the first attempt doesn't resolve the issue, try a solution of your own based on the suggested solution
4. If problems persist, start a fresh conversation and seed it with any essential data from the broken conversation

> ***Remember:*** *Starting a new conversation is often the most efficient solution when quick fixes don't work. There's no penalty for restarting, and a fresh context frequently leads to better results than continuing with a problematic exchange.*

Basic Response Issues

Situation	Do This
AI repeating itself	"Let's pause. The key question is [restate the main point]"
Made-up statistics	"Please only include verifiable facts or provide general ranges"
Conflicting statements	Quote both contradictions and request explicit correction
Declining coherence	Break the next request into numbered steps (max 3) or fork to a new conversation
Ignored requirements	Bullet point all skipped items and request specific coverage
Vague responses	Request particular examples for each point
Over-generalization	Ask for three specific instances
Excessive detail	"Let's focus on the core points: [list them]"
Tangential responses	"Returning to the main question: [restate]"
Missing context	State three key background facts

Table 4.7.1: AI Troubleshooting Matrix - Basic Response Issues

Memory Management

Situation	Do This
Forgetting earlier info	Quote the specific forgotten point
Mixing up projects	"We're focusing only on [project name + one-line description]"
Lost constraints	List all constraints numbered, request confirmation
Confused chronology	Create three-point timeline
Task amnesia	Restate single clear objective
Context bleed	"Starting fresh with only [current topic]"
Parameter drift	Restate original parameters verbatim
Reference confusion	Clarify specific source or reference
Lost personalization	Restate key personal/project details
History confusion	Reset with current task only

Table 4.7.2: AI Troubleshooting Matrix - Memory Management

Format Control

Situation	Do This
Wrong output format	Show exact template with placeholders
Skipped sections	Create a checklist of required sections
Mixed formats	Provide single format example
Structure breakdown	Give numbered outline
Style inconsistency	State three key style rules
Wrong sectioning	Provide section headers in order
Poor organization	Give a clear hierarchy of points
Format mixing	Show correct format once, request adherence
Layout issues	Provide spatial organization rules
Inconsistent notation	Define exact notation system

Table 4.7.3: AI Troubleshooting Matrix - Format Control

Technical Tasks

Situation	Do This
Incorrect syntax	Show minimal working example
Function scope creep	List only required functions with single-line descriptions
Missing error handling	List three critical error cases
Poor documentation	Provide documentation template
Integration issues	List exact dependencies and versions
API confusion	Show exact API call format
Parameter misuse	Define parameter types and ranges
Library confusion	Specify exact library and version
Framework mismatch	State exact framework requirements
Tool misapplication	Define correct tool usage

Table 4.7.4: AI Troubleshooting Matrix - Technical Tasks

Creative Tasks

Situation	Do This
Voice inconsistency	Provide two representative voice examples
Mixed messaging	State single core message
Style drift	Give three defining style characteristics
Weak transitions	Mark exact transition points
Structure issues	Provide skeleton outline
Tone problems	Give two tone examples
Character inconsistency	List key character traits
Plot deviation	Restate core plot points
Pacing issues	Define rhythm pattern
Theme dilution	State primary theme

Table 4.7.5: AI Troubleshooting Matrix - Creative Tasks

Analysis Approaches

Situation	Do This
Incomplete analysis	List minimal required data points
Unsupported conclusions	Request specific evidence for each point
Scope wandering	State single analysis question
Poor comparisons	Define exact comparison parameters
Missing metrics	List required metrics
Methodology issues	State required analytical method
Invalid assumptions	List core assumptions for verification
Framework misapplication	Define correct framework usage
Data gaps	Specify missing data requirements
Logic jumps	Request step-by-step reasoning

Table 4.7.6: AI Troubleshooting Matrix - Analysis Approaches

Process Management

Situation	Do This
Workflow confusion	Create numbered step sequence
Resource misallocation	List required resources by step
Timeline issues	Create milestone list
Dependency problems	Show dependency chain
Stakeholder confusion	List key stakeholders and roles
Priority confusion	Create priority hierarchy
Quality issues	State quality criteria
Communication gaps	Define communication points
Responsibility confusion	Create RACI matrix
Deliverable mismatch	List required deliverables

Table 4.7.6: AI Troubleshooting Matrix - Process Management

Recovery Actions

Situation	Do This
Complete breakdown	"Full stop. Starting fresh with: [single goal]"
Circular logic	Provide new sequential framework
Lost context	Quote three most recent key points
Tool failure	Switch to manual step-by-step
System confusion	State three basic system rules
Critical error	Halt and request system check
Data corruption	Return to last valid state
Process failure	Restart with minimal process
Logic collapse	Begin with single premise
Communication breakdown	Reset to basic yes/no questions

Table 4.7.7: AI Troubleshooting Matrix - Recovery Actions

Clarification Requests

Situation	Do This
Ambiguous terms	Request definition of key terms
Unclear objectives	Ask, "What does success look like?"
Vague requirements	Request specific acceptance criteria
Undefined scope	Ask for explicit boundaries
Unclear priorities	Request priority ranking
Ambiguous metrics	Ask for specific measurement criteria
Unclear authority	Request decision-maker identification
Vague timeline	Ask for specific deadlines
Undefined quality	Request quality standards
Unclear dependencies	Ask for explicit dependencies

Table 4.7.8: AI Troubleshooting Matrix - Clarification Requests

4.8 Practical Exercises: Putting It All Together

Now that we've explored these techniques, let's put them into practice with some exercises. Think of these as dance practice sessions with your AI partner:

Exercise 1: The Context Window Dance

Goal: Practice managing the AI's context window.

1. Start a conversation about climate change.
2. After a few exchanges, abruptly change the topic to ancient Egyptian architecture.
3. Then, try to bring the conversation back to climate change without explicitly restating the topic.
4. Observe how the AI handles this context switch. Does it maintain information from both topics? How do you need to guide it to connect these seemingly disparate subjects?

Exercise 2: The Perspective Shift

Goal: Practice guiding the AI to consider multiple perspectives.

1. Ask the AI to explain the pros and cons of online video games from the perspective of a teenager.
2. Then, ask for the perspective of a parent.
3. Finally, ask for the perspective of a social scientist.
4. Compare and contrast these viewpoints. How did your prompting affect the AI's responses?

Exercise 3: The Creativity Challenge

Goal: Push the boundaries of the AI's creative capabilities.

1. Ask the AI to invent a new sport combining chess and basketball elements.
2. Then, ask it to write a short story (about 100 words) about a championship match of this new sport.
3. Finally, ask it to create a logo for this sport, describing the logo in detail.

Reflect on how you needed to guide the AI through this creative process. What techniques were most effective?

4.9 Conclusion: Mastering the Art of AI Conversation

As we wrap up this chapter, remember that conversing with AI is an art form. Like any art, it takes practice, creativity, and a willingness to experiment. You're not just inputting queries and receiving outputs – you're engaging in a dynamic collaboration with a powerful cognitive tool.

The techniques we've explored – crafting effective prompts to guide responses, managing context, and leveraging different knowledge sources – are your toolkit for this collaboration. As you practice and refine these skills, you can tackle increasingly complex tasks and generate more innovative ideas with your AI partner.

Remember, the goal isn't to make the AI think like a human but to leverage its unique capabilities to complement and enhance human thinking. You're the choreographer in this dance of ideas, guiding the AI's vast knowledge and processing power toward your specific goals and insights.

As you continue your journey in AI collaboration, stay curious, keep experimenting, and don't be afraid to push boundaries, within reason of course. The most exciting discoveries often happen when we venture into uncharted territory. Who knows what new insights and innovations you might uncover in your conversations with AI?

In our next chapter, we'll explore how to harness these conversational skills for idea generation, taking our journey with AI into even more creative and productive realms. Get ready to brainstorm like never before!

CHAPTER 4 SUMMARY: CONVERSING WITH AI: TECHNIQUES AND STRATEGIES

Key Concepts:

1. The art of effective prompting and how to craft clear, specific prompts
2. Guiding AI responses through strategic conversation techniques
3. Advanced techniques for complex AI collaboration and problem-solving
4. Troubleshooting techniques using action matrices for common AI interaction issues
5. Methods for extracting maximum value from AI conversations through iteration and refinement

Reflection Questions:

1. How might you use the various prompting techniques to improve your AI interactions?
2. What obstacles do you foresee when introducing multi-step strategies in intricate AI dialogues?
3. How can you effectively use the troubleshooting matrices to improve your AI collaboration outcomes?
4. How might you adapt these techniques to different field AI tasks?

Action Items:

1. Practice crafting structured prompts using the techniques discussed
2. Experiment with the troubleshooting matrices on a current AI project
3. Develop a personal library of effective prompts for everyday tasks
4. Create a workflow that incorporates the push-pull dynamics discussed

Further Exploration:

- Research advanced prompt engineering techniques
- Study successful examples of complex AI conversations
- Explore ways to combine different conversation techniques for better results
- Investigate how different AI models respond to various prompting strategies

5

HARNESSING AI FOR IDEA GENERATION

5.1 Introduction: Unleashing Creative Potential Through AI Collaboration

So far in this book, we've explored the intricacies of chat AIs and discussed the art of conversation with them. Let's embark on a new adventure: harnessing AI for idea generation. This chapter is about transforming raw data and abstract concepts into innovative ideas.

5.2 Brainstorming with AI

Brainstorming takes on new dimensions when paired with artificial intelligence. Here, we'll explore techniques for turning your AI companion into a tireless idea factory capable of generating concepts that range from practical to fantastical.

THE DIVERGENT THINKING techniques we're exploring here leverage the attention-directing principles from Chapter 3.5. For specific applications of these brainstorming methods in professional contexts, see:

- Content creation examples in Section 8.2
- Educational applications in Section 8.3
- Research and development cases in Section 8.5

> *Tip: Advanced users should also reference Chapter 12.3's discussion of how these techniques evolve with next-generation AI systems.*

THE DIVERGENT THINKING TECHNIQUE

To kick off our exploration, let's explore divergent thinking. This approach generates many ideas without immediate judgment, allowing for ad hoc creativity.

Start Broad: Begin with a broad, open-ended prompt. For example: "List 20 unconventional uses for a paperclip in a post-apocalyptic world."

Embrace Quantity: Encourage the AI to generate many ideas. The goal here is volume, not quality - that comes later.

Mix and Match: Ask the AI to combine random ideas from its list to create even more novel concepts.

Flip Perspectives: Request ideas from different viewpoints. "Now, give me ten uses for a paperclip from the perspective of a time-traveling Victorian aristocrat."

The SCAMPER Method

SCAMPER stands for Substitute, Combine, Adapt, Modify, Put to another use, Eliminate, and Reverse. This method can be powerfully augmented with AI assistance.

Substitute: "What if we replace the paperclip with a piece of string? How would that change our list?"

Combine: "Combine the paperclip with one of the other objects you've mentioned. What new uses emerge?"

Adapt: "How could we adapt the paperclip to solve a problem in underwater exploration?"

Modify: "If the paperclip were 100 times larger, what new uses would it have?"

Put to another use: "How could a paperclip be used in a fine dining restaurant?"

Eliminate: "If we remove the paperclip's ability to bend, what uses remain?"

Reverse: "What if the paperclip's purpose was to separate things instead of holding them together?"

By guiding the AI through each step of SCAMPER, you can generate a wealth of innovative ideas and perspectives.

The Analogy Approach

Analogies can be a powerful tool for idea generation, and AI excels at drawing unexpected connections.

Start with a Concept: Choose a concept or problem you're working on.

Generate Analogies: Ask the AI to generate analogies from various fields. "What natural processes are analogous to our product distribution system?"

Explore the Analogies: Ask the AI to explain how each analogy relates to your original concept and what insights can be drawn from it.

Apply Insights: Use these insights to generate new ideas or solutions for your original problem.

> ***Remember:*** *The key to effective brainstorming with AI is to balance guiding the AI's focus and allowing for unexpected connections. By visualizing the AI's attention as a spotlight that you can direct, you can illuminate new corners of the idea landscape, uncovering gems of innovation that might otherwise remain hidden.*

Concrete Examples of AI-Assisted Ideation

Let's explore how AI can assist in ideation across various fields:

Product Design:

You: "Generate ten ideas for eco-friendly packaging alternatives for electronics."

AI Response: "1. Biodegradable mycelium foam packaging, 2. Reusable bamboo cases, 3. Recycled denim insulation wraps..."

Marketing Campaign:

You: "Brainstorm viral marketing ideas for a new plant-based meat alternative."

AI Response: "1. 'Meat the Future' VR farm tour experience, 2. Social media 'Taste Test Challenge' with celebrities, 3. Pop-up 'Future Food' festivals in major cities..."

Urban Planning:

Prompt: "List innovative ideas for transforming unused urban spaces into community assets."

AI Response: "1. Vertical gardens on blank building walls, 2. Micro-parks in parking spaces, 3. Underground mushroom farms in abandoned subway stations..."

Educational Technology:

Prompt: "Generate concepts for AR/VR applications to enhance history education."

AI Response: "1. 'Time Travel Classroom' - immersive historical recreations, 2. 'History Detective' - gamified archaeological simulations, 3. 'Living Paintings' - AR-enhanced museum experiences..."

Sustainable Energy:

> **Prompt:** "Brainstorm unconventional sources of renewable energy for small communities."
>
> **AI Response:** "1. Piezoelectric sidewalks generating electricity from footsteps, 2. Algae-powered streetlights, 3. Wind-harvesting 'tree' sculptures..."

5.3 Structuring Complex Ideas

After ideation, the critical task is transforming your raw concepts into structured, actionable frameworks. This requires divergent thinking (for generating ideas) and convergent thinking (for organizing them).

AI EXCELS AT BOTH THROUGH SPECIFIC TECHNIQUES:

- Clustering related concepts
- Identifying hierarchical relationships
- Mapping conceptual dependencies
- Suggesting organizational schemas based on recognized patterns

Mind Mapping with AI

Mind mapping is a way to organize knowledge or ideas visually. AI can help create rich, complex mind maps.

Start with a Central Concept: Ask the AI to center your main idea or problem on a mind map.

Branch Out: Request the AI to create main branches for key subtopics or categories.

Add Depth: For each branch, ask the AI to add several levels of subtopics, encouraging it to make connections between different branches.

Cross-Pollinate: Challenge the AI to find connections between seemingly unrelated branches, fostering creative links.

Prune and Refine: Ask the AI to identify the most promising or relevant branches and suggest areas for further expansion.

The Eisenhower Matrix

The Eisenhower Matrix (the Urgent-Important Matrix) can be a powerful tool for prioritizing ideas or tasks, significantly enhanced with AI insights.

Categorize: Ask the AI to categorize your ideas or tasks into four quadrants: Urgent and Important, Important but Not Urgent, Urgent but Not Important, and Neither Urgent nor Important.

Analyze: Request the AI to provide a rationale for each categorization, encouraging more profound thought about priorities.

Optimize: Ask the AI to suggest ways to reduce items in the "Urgent but Not Important" category, freeing up more time for essential tasks.

Action Plan: Have the AI create a draft action plan based on the matrix, suggesting how to tackle each quadrant.

Concept Mapping for Complex Systems

When dealing with intricate systems or multifaceted problems, concept mapping can help clarify relationships and dependencies.

Identify Key Concepts: Ask the AI to list the main concepts or components of your complex idea or system.

Establish Relationships: Ask the AI to connect these concepts, specifying how they relate. For example:

- Identify which concepts influence others
- Determine if some concepts are subsets of more enormous ideas
- Highlight cause-and-effect relationships

To visualize these connections, you could request the AI to represent the relationships as a Directed Acyclic Graph (DAG). Once created, ask the AI to perform a topological sort of the DAG, which can reveal the logical order or dependencies between concepts.

Identify Feedback Loops: Challenge the AI to find circular relationships or feedback loops within the system.

Analyze Impacts: Ask the AI to trace its impact through the system for each concept, identifying key leverage points and potential unintended consequences.

Simplify: Request the AI to suggest ways to simplify the system without losing essential functionality.

By leveraging these AI-assisted structuring techniques, you can transform a chaotic collection of ideas into organized, actionable, and more presentable frameworks. The key is continually guiding the AI's attention, asking it to focus on different aspects of the structure and make connections that might not be immediately obvious.

5.4 Enhancing Problem-Solving Skills

With a wealth of ideas and structured frameworks, we can now focus on problem-solving. AI can be an exceptional partner in this process, offering novel perspectives and rigorous analytical capabilities. This is particularly true when tackling coding challenges and comparing complex options.

Leveraging AI for Coding Problems

When dealing with coding issues, AI can be an invaluable assistant. Here are some techniques to enhance your problem-solving approach:

Chain of Thought (COT) Reasoning: Ask the AI to use COT reasoning when analyzing code.

> "Please use Chain of Thought (COT) reasoning to explain how this function works, step by step."

Root Cause Analysis (RCA): Request the AI to perform RCA on code errors.

> "Please perform a root cause analysis on why this code is producing unexpected output."

Code Tracing: Instruct the AI to trace through the code it has seen.

> "Please trace through this code until you find the root cause of the error."

STRATEGIC INPUT ORDERING: When providing both code files and error messages, consider the order carefully:

> **Errors before code:** This can help the AI focus immediately on the problem area.
>
> **Code before errors:** This allows the AI to understand the context before analyzing the error.
>
> Experiment with both approaches to see which yields better results for your problem.

COMPARATIVE ANALYSIS TECHNIQUES

When you need to compare different options or features, AI can help create structured comparisons:

Scorecard Development: Ask the AI to develop a scorecard for comparison.

> "Please create a scorecard with criteria for evaluating different cloud storage solutions."

Scorecard Population: Once you have a scorecard, ask the AI to populate it with sample cases.

> "Using the scorecard we developed, can you evaluate Google Drive, Dropbox, and OneDrive?"

Feature Comparison Tables: Request the AI to create tables for detailed comparisons.

> "Please create a feature comparison table for the top 5 JavaScript frameworks, including factors like performance, community support, and learning curve."

Example response:

Framework	Performance	Community Support	Learning Curve
React	High	Excellent	Moderate
Vue	High	Good	Easy
Angular	High	Very Good	Steep
Svelte	Very High	Growing	Easy
Ember	Good	Moderate	Moderate

Table 5.4.1: Feature Comparison Table - Example

The Five Whys Technique

Initially developed by Sakichi Toyoda for Toyota, the Five Whys technique can be powerfully augmented with AI assistance.

State the Problem: Begin by clearly stating the problem to the AI.

Ask Why: Have the AI ask "Why?" and provide a possible answer.

Repeat: Continue this process at least five times, with each "Why?" delving deeper into the root cause.

Branch Out: If multiple causes emerge, ask the AI to create separate "Why?" chains for each.

Analyze: Request the AI to analyze the final "Why?" answers and suggest potential solutions addressing these root causes.

TRIZ (Theory of Inventive Problem Solving)

TRIZ, which stands for "Teoriya Resheniya Izobretatelskikh Zadatch" in Russian (Theory of Inventive Problem Solving), is a robust methodology for systematic innovation. Developed by Genrich Altshuller and his colleagues in the Soviet Union beginning in the 1940s, TRIZ offers a structured approach to problem-solving that relies on logic, data, and research rather than intuition or trial-and-error.

Critical concepts in TRIZ include:

Contradictions: Many problems arise when improving one aspect of a system worsens another.

40 Inventive Principles: Altshuller identified 40 principles inventors used to resolve contradictions.

Ideality: TRIZ encourages thinking about the "ideal final result" - a perfect solution that achieves the desired function without cost or harm.

The TRIZ process, when combined with AI assistance, can be compelling:

Define the Problem: Clearly articulate the problem to the AI. For example, "Design a coffee mug that keeps beverages hot for longer without burning the user's hands."

Identify Contradictions: Ask the AI to identify inherent contradictions in the problem. In our example, the mug needs to retain heat for the beverage while remaining cool to the touch for the user's comfort.

Apply TRIZ Principles: Request the AI to suggest relevant TRIZ principles to resolve these contradictions. For instance, "How might we apply the 'Another Dimension' principle to solve this coffee mug problem?"

Generate Solutions: Have the AI generate multiple solution concepts based on these principles. Using the 'Another Dimension' principle, the AI might suggest a double-walled mug with a vacuum between the walls.

Evaluate and Refine: Ask the AI to evaluate each solution against your criteria and suggest refinements. For example, "How can we improve the insulation properties while keeping manufacturing costs low?"

Benefits of using TRIZ with AI include structured creativity, vast knowledge application, rapid iteration, and overcoming cognitive biases. AI can quickly apply TRIZ principles across domains, drawing from different examples and applications.

By combining TRIZ methodology with AI assistance, innovators can tap into a powerful synthesis of systematic thinking and rapid ideation, potentially leading to breakthrough solutions in various fields. This approach is particularly well-suited for AI assistance due to its structured nature, allowing for efficient problem-solving and creative idea generation.

The systematic nature of TRIZ aligns well with AI's capabilities, making it an excellent tool for enhancing innovative problem-solving across various industries and disciplines. AI is likely only to grow more capable of effectively applying TRIZ principles, and its ability to generate novel solutions is also expected to improve, further amplifying the potential of this robust problem-solving methodology.

Lateral Thinking Techniques

Lateral thinking refers to a method of problem-solving that approaches challenges from unexpected angles. Edward de Bono coined the concept and its name in 1967. Unlike vertical thinking, which follows a step-by-step logical progression, lateral thinking involves looking at problems from fresh perspectives to generate creative solutions. Combined with AI assistance, these techniques can lead to innovative outcomes.

Lateral thinking techniques particularly enhanced by AI include:

Random Entry: This technique introduces an unrelated word or concept to stimulate new ideas. With AI:

- Ask the AI to generate a genuinely random word or concept, or you may choose to supply one of your own.
- Challenge the AI to connect this random element to your problem meaningfully.
- For example, if working on improving urban transportation and the AI suggests a "butterfly," it might lead to ideas about lightweight, flexible vehicles or a system that adapts to changing urban landscapes.

Provocation: This method uses provocative statements to challenge existing assumptions. With AI:

- Have the AI create a provocative or outrageous statement about your problem.
- Explore the consequences and implications of this provocation.
- For example, for a retail business problem, the AI might propose: "What if we had no physical products?" This could lead to innovative ideas about virtual goods, service-based models, or experiential retail.

Reverse Thinking: This involves completely reversing the problem or goal to gain new insights. With AI:

- Ask the AI to formulate the exact opposite of your current problem or goal.
- Work backward from this reversal to generate new perspectives on the original problem.
- For example, the AI might explore "How can we increase customer retention?" instead of "How can we ensure customers never come back?" This reversal could reveal overlooked aspects of customer experience.

Analogy Generation: This technique compares unrelated fields to inspire new ideas. With AI:

- Request the AI to generate analogies from diverse, seemingly unrelated fields.
- Apply these analogies to your problem to uncover novel approaches.
- For example, when tackling a supply chain issue, the AI might draw an analogy with the human circulatory system, potentially leading to ideas about adaptive, self-regulating distribution networks.

BENEFITS OF USING lateral thinking techniques with AI include:

Expanded Idea Generation: AI can rapidly produce a vast array of random entries, provocations, reversals, and analogies, significantly broadening the scope of potential solutions.

Overcoming Mental Blocks: AI may not suffer from the same cognitive biases or fixed thinking patterns that might have triggered your mental block, potentially leading to more diverse and unexpected connections.

Rapid Iteration: AI can quickly generate and iterate on ideas, allowing for faster exploration of unconventional solution spaces.

Cross-Domain Insights: AI's broad knowledge base enables it to draw analogies and connections from various fields, potentially leading to cross-pollination of ideas.

When applying these techniques, it's essential to:

- Remain open to seemingly absurd or impossible ideas, which may lead to practical innovations.
- Use AI-generated ideas as springboards for human creativity rather than as final solutions.
- Combine multiple techniques for more comprehensive problem-solving.

By leveraging AI to enhance lateral thinking techniques, problem-solvers can break free from conventional thought patterns and explore truly innovative solution spaces. This approach can be precious in fields requiring breakthrough thinking or when traditional methods are limited.

THE CYNEFIN FRAMEWORK

The Cynefin Framework, developed by Dave Snowden in 1999, is a tool that helps leaders understand their situation better, empowering them to make wise, informed decisions. This framework is beneficial for dealing with complex, ambiguous problems where the relationship between cause and effect is soon apparent. The Cynefin Framework can become an even more powerful tool for navigating uncertainty and complexity when combined with "A.I." assistance.

The framework consists of five distinct domains, each needing its own approach to problem-solving.

Let's explore how to use it effectively:

First, you need to categorize your situation. When working with AI assistance, start by describing your problem in detail. Ask the AI to analyze the characteristics and suggest which domain fits best. For instance, if you're trying to predict how customers will react in a new market, the AI might identify this as a Complex situation because it involves many unknown factors and unpredictable interactions.

Now, let's explore each domain:

The first domain is **Clear**, which was previously known as Simple or Obvious. In Clear situations, you can easily see how cause leads to effect. The approach here is straightforward: Sense what's happening, Categorize it, and Respond with best practices. AI can help by quickly spotting patterns and automating routine decisions.

The second domain is **Complicated**. Here, cause and effect are connected, but not obviously so. The approach is to Sense the situation, Analyze it carefully, and then Respond. Think of it like solving a puzzle - it takes expertise but has a solution. AI excels here by analyzing large amounts of data and finding hidden patterns.

Third is the **Complex** domain. This is where cause and effect only become clear after the fact, and many elements interact with each other. The approach is to Probe with experiments, Sense what happens, and then Respond. AI can help by suggesting different possibilities to test and analyzing the results of small experiments.

Fourth is the **Chaotic** domain. In chaos, there are no clear patterns - think of crisis situations. The approach is to Act first to establish some order, then Sense what's working, and Respond accordingly. AI can be valuable here by quickly processing real-time information and suggesting immediate actions.

Finally, there's **Disorder**. This is when you're not sure which of the other domains applies. The goal is to gather enough information to move the situation into one of the other domains. AI can help analyze what you know so far and suggest what other information you need to collect.

The beauty of this framework is that once you understand which domain you're in, you can choose the right approach and tools for the situation at hand. By combining human wisdom with AI capabilities, leaders can navigate even the most challenging scenarios with greater confidence and effectiveness.

USING The Cynefin Framework's Probe-Sense-Respond process for Complex Problems: In complex domains, the Probe-Sense-Respond approach is precious. With AI:

Probing: Ask the AI to generate multiple small, safe-to-fail experiments to test hypotheses about the complex system. Example: For a complex market entry problem, the AI might suggest A/B testing different marketing messages in small, diverse customer segments.

Sensing: Use AI to analyze the results of these experiments, identifying emerging patterns and potential attractors. Example: The AI could process customer response data, social media reactions, and sales figures to identify which approaches are gaining traction.

Responding: Based on the analysis, have the AI suggest adaptive responses and next steps. Example: The AI might recommend scaling up successful approaches, modifying less successful ones, or exploring new directions based on unexpected results.

Benefits of using the Cynefin Framework with AI include:

Enhanced Situational Awareness: AI can process vast amounts of data to help accurately categorize problems and identify shifts between domains.

Rapid Experimentation: AI can quickly generate and analyze numerous safe-to-fail experiments in complex domains, accelerating learning.

Pattern Recognition: AI excels at recognizing subtle patterns in complex systems, aiding in sense-making.

Adaptive Decision Support: AI can continuously update its recommendations as situations evolve, supporting agile decision-making.

When applying the Cynefin Framework with AI, it's essential to:

- Regularly reassess the problem's domain, as situations can shift between domains over time.
- Use AI insights as input for human decision-making, especially in Complex and Chaotic domains where human judgment and creativity are crucial.
- Maintain a balance between exploiting AI-identified patterns and exploring new possibilities to avoid premature convergence on suboptimal solutions.

By leveraging AI with the Cynefin Framework, decision-makers can navigate complex and uncertain environments more effectively, making informed choices based on a nuanced understanding of the problem context.

A Note on Mathematical Problem-Solving

It's crucial to understand the limitations of Language Model AIs when it comes to mathematical computations:

> **Limitations:** Understand that Language Model AIs have limitations in complex mathematical computations.
>
> **Verification:** Always verify AI-generated mathematical results independently, especially for critical applications.
>
> **Conceptual Understanding:** Focus on using AI for conceptual understanding and problem framing in mathematical contexts rather than for precise calculations.

LLMs do their best at math when they call dependable tools, such as Python, in order to solve equations. Regardless, you should independently verify mathematical conclusions critical to your use cases, as the LLM might also err in how it sets up the problem or it may be mistaken in how it calls external tools.

5.5 Combining Human Creativity with AI Suggestions

While AI can generate many ideas, the true magic happens when human creativity intertwines with AI suggestions. Here are some techniques to foster this magic:

Idea Ping-Pong: Start with a human-generated idea, then ask the AI to build upon it. Alternate between human and AI contributions, each time pushing the concept further.

AI as Devil's Advocate: After generating your ideas, ask the AI to evaluate them, critically pointing out potential flaws or limitations. This can help refine and strengthen your concepts.

Fusion Thinking: Combine seemingly unrelated ideas—one from a human and one from an AI—to create novel concepts. For example, "How can we combine the sharing economy with space exploration?"

Constraint Introduction: Have the AI suggest unusual constraints or limitations, then use your human creativity to solve problems within these boundaries.

Metaphor Expansion: Provide a metaphor for your problem or solution, then ask the AI to expand on it, drawing out insights you can interpret and apply creatively.

5.6 Techniques for Evaluating and Refining AI-Generated Ideas

Not all ideas are created equal. Here's how to separate the wheat from the chaff:

Feasibility Assessment: Ask the AI to rate each idea on a scale of 1-10 for technical feasibility, then have it explain its reasoning. Use this as a starting point for your evaluation.

Impact vs. Effort Matrix: Have the AI place ideas on a 2x2 matrix of impact vs. effort. Focus on high-impact, low-effort ideas first, but keep the high-impact, high-effort ones.

SWOT Analysis: Request a SWOT (Strengths, Weaknesses, Opportunities, Threats) analysis for the most promising ideas. This can highlight potential issues and advantages you might have missed.

Scenario Planning: Consider exploring three possible outcomes for each idea: the best, worst, and most likely. This approach will help you prepare effectively and enhance your plan, setting you up for success no matter the situation.

Stakeholder Impact Assessment: Have the AI consider how different stakeholders might be affected by each idea. This can help you identify potential resistance or support.

Iterative Refinement: Combine the best elements from multiple AI-generated ideas. Then, ask AI to refine this new hybrid concept further.

5.7 Step-by-Step Guide for an AI-Assisted Brainstorming Session

Here's a comprehensive guide to conducting an effective AI-assisted brainstorming session:

Prepare the Problem Statement: Clearly define the problem or goal you're addressing.

> "We're brainstorming solutions to reduce urban traffic congestion."

Set the Stage: Explain the brainstorming rules to the AI (e.g., no idea is too wild, quantity over quality initially).

> "In this session, we want to generate as many ideas as possible without judging them initially."

Initial Idea Generation: Ask the AI to generate a large number of ideas.

> "Please generate 50 ideas to reduce urban traffic congestion."

Human Contribution: Review the AI-generated ideas and add your own.

> "I've thought of a few more ideas. Let's add them to the list."

Categorization: Ask the AI to categorize the ideas into themes or approaches.

> "Can you group these ideas into 5-7 main categories or themes?"

Expansion and Combination: Select promising ideas or categories and ask the AI to expound.

> "Let's take the 'Smart Traffic Management' category. Can you generate ten more specific ideas within this theme?"

Wild Card Round: Introduce a random element to spark new thinking.

> "Now, let's introduce the concept of 'underwater' to our traffic problem. How might this lead to new solutions?"

Critical Evaluation: Begin to filter ideas based on feasibility and impact.

> "For each of our top 10 ideas, can you provide a brief SWOT analysis?"

Refinement: Take the best elements from multiple ideas and combine them.

> "Let's combine aspects of ideas 3, 7, and 12. How might this new hybrid solution work?"

Action Planning: Ask the AI to outline the next steps for the most promising ideas.

> "For our top 3 ideas, can you create a high-level implementation plan?"

Reflection: Ask the AI to summarize key insights from the session.

> "What are the main themes or patterns you've noticed in our brainstorming session?"

Follow-up: Set tasks for further research or development.

> "Based on our session, what are the three most critical areas we should research further?"

Remember, this is an iterative process. If new insights emerge or you feel certain areas need more exploration, don't hesitate to loop back to earlier steps.

5.8 Iterative Refinement and Validation

The process of idea generation and problem-solving with AI is inherently iterative. Here's how to refine and validate your solutions:

Peer Review Simulation: Ask the AI to take on the role of different experts to critique the solution.

> "As an experienced software architect, what potential issues do you see with this design?"

Stress Testing: Challenge the AI to find edge cases or potential failure points.

> "What are the extreme conditions under which this solution might fail?"

Implementation Planning: Have the AI help break down the solution into actionable steps.

> "Please create a high-level implementation plan for this solution, including potential challenges at each stage."

Ethical Consideration: Always ask the AI to evaluate potential ethical implications.

> "What ethical concerns should we consider with this solution, and how might we address them?"

Feedback Loop: Use the insights from these validation steps to refine your original ideas, creating a continuous improvement cycle.

Comparative Analysis: Ask the AI to compare your refined solution with existing alternatives.

> "How does our solution compare efficiency, cost, and user experience to current market leaders?"

Future-Proofing: Challenge the AI to consider future trends that might impact your solution.

> "Given current technological trends, how might this solution need to evolve over the next 5-10 years?"

5.9 Case Studies: AI-Assisted Ideation Success Stories

To illustrate the power of AI-assisted ideation, let's look at a few hypothetical case studies:

Sustainable Packaging Innovation: A consumer goods company used AI-assisted brainstorming to reimagine its packaging. The AI suggested biomimicry approaches, leading to a new packaging design inspired by honeycomb structures. This resulted in a 40% reduction in material use while improving product protection.

Urban Mobility Solution: A city planning team used AI to generate ideas for reducing traffic congestion. The AI proposed a dynamic road pricing system based on real-time traffic data and an AI-powered public transport routing app. Implementation led to a 30% reduction in peak-hour congestion.

Education Technology Breakthrough: An edtech startup used AI-assisted ideation to develop a new learning platform. The AI suggested combining spaced repetition techniques with virtual reality, leading to an immersive language learning app that improved retention rates by 60% compared to traditional methods.

Healthcare Diagnostic Tool: A medical research team used AI to brainstorm new approaches to early disease detection. This led to the invention of a multi-biomarker blood test combined with machine learning analysis, resulting in a screening tool that could detect multiple types of cancer at early stages.

These case studies demonstrate how AI-assisted ideation might lead to innovative solutions across various fields, often by combining ideas unexpectedly or applying concepts from one domain to another.

5.10 Conclusion: From Ideation to Innovation

Reflect on how we've transformed AI from a tool to an active partner in ideation, organization, and problem-solving. By skillfully directing the AI's attention and leveraging its unique capabilities, we've unlocked new realms of creativity and innovation.

The key to successful AI-assisted ideation is balancing human creativity and AI capabilities. While AI can generate many ideas and make connections that humans might miss, the human touch provides context, judges relevance, and drives implementation.

Remember that AI-assisted ideation is not about replacing human creativity but augmenting it. The most potent solutions emerge when we combine the breadth and speed of AI-generated ideas with human insight, experience, and intuition.

As you move forward, continue to experiment with different techniques and approaches. Each problem may require a unique combination of methods, and your skill in AI-assisted ideation will grow with practice.

In our next chapter, we'll explore how to extract even more value from our AI interactions, turning our conversations into actionable insights and tangible outputs. Until then, keep exploring, questioning, and pushing the boundaries of what's possible in your dance with artificial minds.

CHAPTER 5 SUMMARY: HARNESSING AI FOR IDEA GENERATION

Key Concepts:

1. Divergent Thinking Technique: Using AI to generate many ideas without immediate judgment.
2. SCAMPER Method: Applying AI to Substitute, Combine, Adapt, Modify, Put to another use, Eliminate, and Reverse ideas.
3. Comparative Analysis Techniques: Leveraging AI for structured comparisons and evaluations of different options or features.
4. Lateral Thinking Techniques: Using AI to approach problems from unexpected angles and generate creative solutions.

Reflection Questions:

1. How might you apply the Divergent Thinking Technique in your next brainstorming session?
2. What challenges do you foresee in applying the SCAMPER method with AI in your field?
3. How does the ethical consideration of potential AI bias impact your approach to using AI for idea generation?

Action Items:

1. Conduct a brainstorming session using the Divergent Thinking Technique with an AI model.
2. Apply the SCAMPER method to your industry's existing product or service.
3. Develop a structured comparison using AI to evaluate potential solutions to a current challenge in your work.

Further Exploration:

- Explore Edward de Bono's "Six Thinking Hats" method and consider how it could be integrated with AI-assisted brainstorming.
- Research advanced prompt engineering techniques for more effective idea generation with AI.
- Investigate the potential long-term implications of relying on AI for creative processes and innovation.

6

EXTRACTING VALUE: BYPRODUCTS OF AI CONVERSATIONS

6.1 Introduction: Mining Gold from Digital Conversations

In our journey, we've explored the intricacies of AI, pursued the art of AI conversation, and harnessed AI for idea generation. Now, we focus on a crucial aspect of AI interaction: extracting tangible value from our digital dialogues. In this chapter, we'll transform AI conversations into practical, actionable outcomes.

6.2 Framework for Identifying Valuable Byproducts

Before we dive into specific techniques, let's establish a framework for identifying valuable byproducts in AI conversations.

This framework will help you systematically extract value from your interactions:

> **Relevance**: How closely does the byproduct align with your initial goals or questions?
>
> **Novelty**: Does the byproduct offer new insights or perspectives?
>
> **Actionability**: Can the byproduct be easily translated into concrete actions or decisions?
>
> **Scalability**: Can the byproduct be applied broadly or scaled to other contexts?
>
> **Reliability**: How confident are you in the accuracy or validity of the byproduct?

When evaluating the outputs of your AI conversations, consider these five factors. Byproducts scoring highly across multiple factors are likely the most valuable.

6.3 Techniques for Extracting Value

Generating Finely-Grained To-Do Lists

One of the most immediate and practical applications of AI conversations is the generation of detailed, actionable to-do lists. When crafted skillfully, AI-generated to-do lists can become comprehensive project plans or intricate roadmaps for personal development.

Techniques for Effective To-Do List Generation:

Brain Dump: Ask the AI to generate a detailed list of tasks for your project or goal.

> "Please create a detailed list of tasks necessary to launch a successful podcast based on the attached brain dump."

Prioritization: Once you have a list, ask the AI to help prioritize.

> "Now, categorize these tasks into 'Must Do,' 'Should Do,' and 'Nice to Have' categories."

Time Estimation: Request time estimates for each task.

> "Provide an estimated completion time for each 'Must Do' task."

Dependencies: Identify task dependencies.

> "Please rank the tasks in the order they should be completed. Please create a dependency chart."

Resource Allocation: If working with a team, ask the AI to suggest resource allocation.

> "Based on these tasks, how would you allocate them among a team of three people?"

Milestones: Have the AI suggest vital milestones.

> "What are the major milestones in this project, and which tasks contribute to each?"

Risk Assessment: Remember potential roadblocks.

> "What are potential risks or challenges for each major task, and how might we mitigate them?"

Analyzing Content and Extracting Insights

A significant advantage of AI-driven discussions is the capability to swiftly analyze and derive insights from diverse content formats, such as articles, video transcripts, and potentially even entire books, provided the AI has a sufficiently sizeable contextual framework.

Techniques for Content Analysis:

Summarization: Ask the AI to summarize the main points.

> "Please summarize the key ideas from this YouTube transcript in 3-5 bullet points."

Key Quote Extraction: Request the AI to identify and extract notable quotes.

> "What are the three most impactful quotes from this content, and why are they significant?"

Thematic Analysis: Have the AI identify recurring themes or concepts.

> "What main themes are discussed in this video, and how do they interconnect?"

Comparative Analysis: If analyzing multiple pieces of content, ask the AI to compare and contrast them.

> "How do the viewpoints in Video A and Video B differ on artificial intelligence?"

Action Item Extraction: Ask the AI to create a list of actionable steps for instructional content.

> "Based on this TED talk, what are the five key actions the speaker recommends for improving productivity?"

Bias Identification: Request the AI to analyze potential biases in the content.

> "Can you identify any potential biases in the speaker's argument, and how might these affect the overall message?"

Further Research Suggestions: Ask the AI to suggest related topics for deeper exploration.

> "Based on the concepts discussed in this video, what are three related topics that would be valuable to research further?"

Creating Knowledge Bases and Summaries

One of the most powerful ways to extract value from AI conversations is to use them to build comprehensive knowledge bases and create detailed summaries of complex topics. This can be particularly useful for research, study, or educational materials.

Techniques for Knowledge Base Creation:

Topic Exploration: Start with a broad topic and ask the AI to break it down into subtopics.

> "Please break down the field of machine learning into its main subfields and key concepts."

Definitions and Explanations: Ask the AI to provide definitions and explanations for each key concept.

> "Can you define 'neural networks' and explain how they work in simple terms?"

Historical Context: Request information about the historical development of ideas or technologies.

> "What is the history of neural networks in AI research?"

Current State of the Art: Ask about the most recent developments in the field.

> "What are the current state-of-the-art applications of neural networks?"

Pros and Cons: Ask the AI to make a bullet point list of the benefits and weaknesses of each concept or technology.

> "What are the pros and cons of using neural networks compared to other machine learning approaches?"

Real-World Applications: Request examples of how concepts are applied in practice.

> "Can you provide three examples of how neural networks are used in industry today?"

Interconnections: Ask the AI to explain how different concepts relate.

> "How do neural networks relate to deep learning and artificial intelligence?"

Future Directions: Inquire about potential future developments.

> "What are some promising future directions for neural network research?"

Learning Resources: Ask for recommendations on how to learn more.

> "What are some good resources (books, courses, websites) for someone wanting to learn more about neural networks?"

6.4 Case Studies of Successful Value Extraction

To demonstrate the practical use of these methods, let's examine four fictional scenarios in which value has been successfully derived from AI conversations.

Case Study 1: Streamlining Product Development

Sarah, a product manager at a tech startup, used AI to streamline her team's product development process. Here's how she extracted value:

Initial Conversation: Sarah had a lengthy conversation with an AI about agile product development methodologies.

To-Do List Generation: She asked the AI to create a detailed to-do list to help her team better implement Scrum.

Resource Allocation: The manager assigned tasks to team members by considering their individual expertise and workload, utilizing AI tools to enhance the efficiency of these assignments.

Risk Assessment: Sarah used the AI to identify potential roadblocks and develop mitigation strategies.

Knowledge Base Creation: She compiled the AI's insights into a comprehensive guide for her team, which became a valuable resource for onboarding new members.

Outcome: Sarah's team adopted Scrum practices more quickly and efficiently, reducing their product development cycle by 30%.

Case Study 2: Enhancing Academic Research

Dr. Chen, a biology researcher, leveraged AI to accelerate his literature review process:

Content Analysis: He used AI to analyze hundreds of research papers on gene editing techniques.

Thematic Analysis: The AI identified key themes and trends across the literature.

Comparative Analysis: Dr. Chen asked the AI to compare different gene editing methods discussed in the papers.

Knowledge Base Development: Leveraging the AI's insights, he constructed a comprehensive database encompassing gene editing techniques, their diverse applications, and potential limitations. He actively participated as the human-in-the-loop to ensure accuracy and relevance, meticulously validating and refining the information.

Outcome: Dr. Chen's literature review was completed in half the usual time, and his resulting paper was more comprehensive and insightful.

Case Study 3: Improving Customer Service Training

Maria, a customer service manager, used AI to enhance her team's training program:

Content Analysis: She fed anonymized transcripts of customer interactions into an AI for analysis.

Key Quote Extraction: The AI identified particularly successful (and unsuccessful) responses to common customer issues.

Action Item Extraction: Maria requested that the AI compile a list of best practices derived from the reviewed interactions.

To-Do List Generation: The AI helped create a detailed plan for implementing these best practices in the team's training program.

Outcome: Customer satisfaction scores increased by 25% after implementing the new training program.

Case Study 4: Personal Skill Development

Alex, a software developer, used AI to guide his learning of a new programming language:

Topic Exploration: He asked the AI to break down the new language's fundamental concepts and explain several code snippets from existing projects.

Learning Resource Recommendations: The AI suggested books, online courses, and coding challenges tailored to Alex's learning style.

To-Do List Generation: Alex used AI to create a structured learning plan with daily coding exercises.

Knowledge Base Creation: As he learned, Alex used AI to help him create a personal reference guide for the new language.

Outcome: Alex became proficient in the new language in half the time he initially estimated, and his reference guide became a valuable resource for his colleagues.

6.5 Challenges and Solutions

While extracting value from AI conversations can be incredibly beneficial, it's challenging. Here are some common issues you might encounter and strategies to overcome them:

Information Overload

Challenge: AI can generate vast amounts of information, which can be overwhelming.

Solution: Use the prioritization techniques discussed earlier. Ask the AI to rank information by importance or relevance. Create summary artifacts for lengthy outputs.

Reliability of Information

Challenge: AI can sometimes provide inaccurate, outdated, or incomplete information.

Solution: Always cross-verify important information from reliable sources. Use the AI's outputs not as the final word but rather as a starting point. Ask the AI to review and improve upon its output.

Context Retention

Challenge: The AI might lose track of earlier context in long conversations.

Solution: Periodically summarize key points and restate important context. Use techniques like "Let's recap what we've discussed so far..."

Bias in AI Outputs

Challenge: AI can reflect and amplify biases in its training data.

Solution: Be aware of potential biases. Ask the AI to consider multiple perspectives on a topic. Use diverse sources to complement AI insights.

Difficulty in Extracting Actionable Insights

Challenge: Sometimes, AI outputs can be too theoretical or abstract.

Solution: Specifically ask for concrete, actionable items. Use prompts like "Based on this information, what are the next three specific actions I can take?"

Integration with Existing Workflows

Challenge: Incorporating AI-generated insights into established processes can be difficult.

Solution: Begin by incorporating AI insights into a single area of your workflow. As you and your team grow more at ease with the process, progressively broaden your use of these tools.

Overreliance on AI

Challenge: There's a risk of becoming too dependent on AI for decision-making.

Solution: Use AI to augment, not replace, human judgment. Regularly reflect on how you're using AI and ensure you're maintaining your critical thinking skills. Ensure that you base your views on more than just discussions with AI.

6.6 Organizing and Storing Valuable Insights

Extracting value from AI conversations is only half the battle; organizing and storing these insights for future use is equally important.

Here are some tips to help you manage the wealth of information you'll generate:

Use a Digital Note-Taking System: Tools like GitHub Gist, Apple Notes, Evernote, Notion, or Obsidian are great for organizing complex information. Create a dedicated space for AI-generated insights. Consider keeping your notes in encrypted storage such as Apple Journal for Intellectual Property such as patent ideas.

Implement a Tagging System: Use tags to categorize insights by topic, project, or application. This makes it easier to find relevant information later.

Create Summary Documents: For each significant AI conversation or analysis, create a summary document highlighting key points, actionable items, and areas for further exploration.

Use Mind Mapping Tools: Tools like MindMeister or XMind can help you visualize connections between concepts and insights.

Set Up a Review System: Regularly review and update your stored insights. Set reminders to revisit important information and reassess its relevance.

Collaborate and Share: When working in a team, use collaborative tools to share and build upon AI-generated insights. This can lead to even more valuable outcomes.

Version Control: For evolving topics, use version control such as git and GitHub.com to track how your understanding and insights change over time.

Create an Insight Journal: Keep a running log of crucial insights, noting when and how you obtained them and any actions taken.

Leverage AI for Insight Organization: Harness AI capabilities to organize and synthesize your collected insights systematically. This creates a higher-level analysis, offering fresh perspectives on your accumulated knowledge. The tech industry is exploring the development of apps that could securely integrate AI with personal note repositories, potentially establishing a new category of productivity tools.

By implementing these organizational strategies, you'll create a valuable, ever-growing repository of knowledge and insights that can inform your decision-making and spark new ideas for years. As new tools arrive, you can migrate your knowledge base into them.

6.7 Conclusion: The Art of Value Extraction

The true power of AI collaboration lies not just in the immediate outputs of our conversations but in our ability to extract, refine, and apply the valuable byproducts of these interactions. By mastering the techniques we've explored – from generating detailed to-do lists to creating comprehensive knowledge bases – you can transform your AI conversations from interesting chats into wellsprings of actionable insights and knowledge.

Remember, the key to successful value extraction is active engagement. Approach each AI interaction with a clear goal, but remain open to unexpected insights. Use the framework we've discussed to identify valuable byproducts, be bold, and ask follow-up questions or request clarifications. With practice, you'll develop an intuitive sense of where each conversation's real value lies.

As you apply these techniques in your work and life, you'll likely discover new and innovative ways to extract value from AI interactions. The examples and case studies we've explored are just the beginning. The field of AI is rapidly evolving, and with it, there is the potential for extracting ever more valuable insights from our digital collaborators.

Next, we'll explore Claude's Artifacts feature. This tool opens up new possibilities for managing and shaping conversation outputs, furthering your AI collaborations. Until then, keep exploring and pushing the limits of what you can achieve with AI.

CHAPTER 6 SUMMARY: EXTRACTING VALUE: BYPRODUCTS OF AI CONVERSATIONS

Key Concepts:

1. Framework for identifying valuable byproducts in AI conversations
2. Techniques for generating finely-grained to-do lists and analyzing content
3. Creating knowledge bases and summaries from AI interactions
4. Overcoming challenges in extracting value from AI conversations

Reflection Questions:

1. How might you apply the framework for identifying valuable byproducts in your following AI conversation?
2. What obstacles do you anticipate when applying content analysis techniques?
3. How does the need for data privacy impact your approach to creating knowledge bases from AI interactions?

Action Items:

1. Practice generating a detailed to-do list using AI assistance for an upcoming project
2. Experiment with using AI to analyze and extract insights from a recent document or article
3. Develop a system for organizing and storing valuable insights gained from AI conversations

Further Exploration:

- Explore tools for visualizing and mapping concepts extracted from AI conversations
- Research advanced techniques for query optimization in retrieval-augmented generation
- Consider the ethical implications of long-term storage and use of AI-generated content

7

MASTERING CLAUDE'S ARTIFACTS: ELEVATING HUMAN-AI COLLABORATION

7.1 Introduction: The Evolution of AI Interaction

Throughout our journey in AI collaboration, we've explored the intricacies of AI cognition and mastered the art of conversation. Now, we stand at the threshold of a paradigm shift in human-AI interaction: Claude's Artifacts feature. This chapter will delve into how Artifacts transform fleeting exchanges into powerful, evolving collaborations, revolutionizing our work with AI.

Artifacts represent a quantum leap in AI interaction, serving as persistent, editable content that transcends the limitations of traditional conversational AI. Imagine a digital ecosystem where your ideas, code, and complex projects persist and evolve through collaborative refinement with AI. This is the power of Artifacts.

THIS EXPLORATION of Claude's Artifacts feature is an evolution of the core concepts we've explored throughout this book:

- Claude Artifact build upon the attention mechanisms discussed in Chapter 3
- They enhance the prompting techniques from Chapter 4
- They streamline the brainstorming processes covered in Chapter 5
- They address many of the organizational challenges outlined in Chapter 9

For practical applications of Artifacts in different fields, refer to the relevant sections of Chapter 8:

Content creation: Section 8.2

Education: Section 8.3

Business: Section 8.4

7.2 Understanding the Artifact Paradigm

The Artifact system is Claude's way of creating and managing substantial, self-contained content within a conversation. But it's much more than just persistent chat bubbles. Let's break down the key components that make Artifacts a game-changer in AI collaboration:

Dynamic Creation: Claude creates Artifacts when the conversation requires substantial, self-contained content, or when you deliberately ask for Artifacts to be created.

Contextual Persistence: Unlike ephemeral chat messages, Artifacts persist across multiple conversational turns, and even conversations at large, maintaining context and allowing for ongoing refinement.

Polymorphic Nature: Artifacts can manifest as code snippets, markdown documents, HTML content, SVG images, and more, adapting to the needs of your project.

Collaborative Editing: Users can request changes, updates, and refinements to existing Artifacts, fostering an iterative, collaborative workflow.

Intelligent Recall: Claude and its users can seamlessly reference and build upon previously created artifacts, creating a web of interconnected knowledge.

Version Control: The system maintains a history of changes, allowing you to track the evolution of ideas and revert if necessary.

Grasping these elements is essential for unlocking Artifacts' full advantages. They go beyond mere features and represent the foundational elements of a groundbreaking approach to AI-enhanced work.

7.3 Practical Applications: Artifacts in Action

The true power of Artifacts lies in their practical applications. Let's explore how they can revolutionize various aspects of your work:

Collaborative Writing and Editing

Artifacts transform the writing process into a dynamic, AI-enhanced experience:

Iterative Drafting: Begin with a high-level outline in your initial Artifact, using clear hierarchical markers (e.g., H1, H2 headers). Ask Claude to refine specific sections with each iteration while maintaining structural integrity. Each version is preserved with a unique identifier, enabling you to track changes, compare versions, and merge elements from different iterations. This versioning system allows for non-destructive editing and comprehensive revision history.

Real-time Collaboration: As you discuss ideas with Claude, relevant portions of the document can be updated in real time, reflecting the flow of your conversation.

Multi-perspective Editing: Ask Claude to review the document from different perspectives (e.g., a subject matter expert, a copy editor, a layperson) to ensure comprehensive refinement.

Version Comparison: Easily compare different versions of your document, with Claude highlighting fundamental changes and their implications.

Advanced Code Development

For developers, Artifacts offers a robust environment for code creation and refinement:

Intelligent Code Generation: Start with a high-level description of your coding task, and watch as Claude generates an initial implementation as an Artifact.

Interactive Debugging: As you encounter issues, Claude can update the code Artifact in real time, explaining each change and its expected impact.

Language Translation: You can easily translate code between programming languages by asking Claude to create a new Artifact based on the original implementation.

Documentation Syncing: Maintain separate but linked Artifacts for code and documentation, ensuring they stay in sync as your project evolves.

Test Suite Generation: Ask Claude to create comprehensive test suites as separate Artifacts based on the functionality described in your code Artifacts.

Dynamic Project Management

Artifacts can act as dynamic records in project management, adapting seamlessly to the changing demands of your project.

Adaptive Planning: Start with a high-level project plan Artifact, then refine and expand it as you discuss details with Claude. The plan evolves with your project, continually reflecting the current and future directions.

Resource Allocation Modeling: Create Artifacts that model different resource allocation scenarios. Claude can help you analyze and optimize these models based on your project constraints.

Risk Assessment Matrix: Develop a dynamic risk assessment Artifact that Claude helps you update based on ongoing discussions about project progress and changing conditions.

Stakeholder Communication: Generate tailored report Artifacts for different stakeholders, pulling relevant information from your master project Artifacts.

7.4 Advanced Techniques: Pushing the Boundaries

As you become more adept with Artifacts, you can employ advanced techniques to enhance your AI collaboration further:

Cross-Pollination of Ideas

Create a network of interconnected knowledge by cross-referencing multiple Artifacts:

Idea Synthesis: Ask Claude to analyze multiple Artifacts and synthesize new insights, creating a new Artifact representing a higher-level understanding.

Contextual Problem-Solving: When facing a challenge, ask Claude to review relevant Artifacts from past conversations, applying learned insights to your current situation.

Interdisciplinary Connections: Explore how concepts from one field (represented in one Artifact) might apply to another, fostering innovative cross-disciplinary solutions.

Artifact-Driven Brainstorming

Use Artifacts to facilitate complex, multi-stage brainstorming sessions:

Idea Expansion: Start with a seed idea Artifact, then ask Claude to generate multiple new Artifacts, each exploring a different aspect or direction.

Conceptual Mapping: Create an Artifact that visually maps the relationships between ideas generated in your brainstorming session.

Scenario Exploration: Develop multiple Artifacts representing different future scenarios, then use these to inform strategy development.

Constraint-Based Ideation: Introduce specific constraints into your Artifact, challenging Claude to generate creative solutions within these boundaries.

Temporal Manipulation

Leverage the persistent nature of Artifacts to manipulate your project's timeline:

Future Projections: Create Artifacts representing future states of your project, then work backward with Claude to determine the steps needed to reach those states.

Historical Analysis: Review the evolution of an Artifact over time, asking Claude to analyze the decision-making process that led to its current state.

Parallel Timelines: Maintain multiple versions of a project Artifact, each representing a different strategic direction, allowing you to compare potential outcomes.

7.5 Ethical Considerations and Best Practices

With great power comes great responsibility. As you leverage Artifacts in your work, keep these ethical considerations in mind:

Data Privacy: Be mindful of sensitive information included in Artifacts. Develop guidelines for what should and shouldn't be included in persistent Artifacts.

Intellectual Property: Establish clear policies on Artifact content ownership and usage rights, especially when collaborating with external parties.

Bias Awareness: Regularly review Artifacts for potential biases, both in content and in the decision-making processes they inform.

Transparency: Maintain precise records of how Artifacts are used in decision-making processes, ensuring accountability and traceability.

Human Oversight: While Artifacts enhance AI collaboration, always apply human judgment to their content and use. Artifacts should augment, not replace, human decision-making.

Ethical Use Guidelines: Develop ethical guidelines for Artifact use within your organization, ensuring all team members understand the responsibilities of this powerful tool.

As AI technology advances, we can look forward to thrilling enhancements in the functionality of artifacts:

Autonomous Artifact Agents: Imagine Artifacts that can update themselves based on new information, alerting you to significant changes or insights.

Cross-Platform Integration: The seamless integration of Artifacts with other productivity tools and AI systems creates a unified digital workspace.

Artifact-Based AI Training: This method would use the history of Artifact interactions to train more personalized AI models, enhancing the AI's understanding of your unique work patterns and needs.

Neural Interface Integration: Direct brain-computer interfaces allowing intuitive, thought-driven Artifact manipulation and creation.

7.6 Conclusion: Artifacts as Extensions of Human Cognition

Claude's Artifacts feature is more than just a tool; it's a new way of thinking about human-AI collaboration. By mastering Artifacts, you're not just improving your productivity – you're expanding the boundaries of your cognitive capabilities.

Artifacts serve as external cognitive scaffolding, allowing you to offload complex information processing and creative tasks to an AI collaborator while maintaining complete control and oversight. They enable you to tackle problems of greater complexity, to see connections that might otherwise remain hidden, and to iterate on ideas with unprecedented speed and insight.

The journey of human-AI collaboration is just beginning, and Artifacts are a glimpse into a future where the lines between human and artificial intelligence blur, creating a symbiosis that amplifies our collective capabilities. Embrace this powerful tool, use it wisely and ethically, and prepare to be amazed at what you can achieve.

In our next chapter, we'll explore how to apply such advanced collaboration techniques across various professional fields, to drive innovation and productivity to new heights.

CHAPTER 7 SUMMARY: MASTERING CLAUDE'S ARTIFACTS: ELEVATING HUMAN-AI COLLABORATION

Key Concepts:

1. Artifacts as persistent, editable content in AI conversations
2. The paradigm shift in AI interaction enabled by Artifacts
3. Practical applications of Artifacts in writing, coding, and project management
4. Advanced techniques for leveraging Artifacts in complex problem-solving
5. Ethical considerations and best practices for Artifact use

Reflection Questions:

1. How might you integrate Artifacts into your current workflow to enhance productivity and creativity?
2. What obstacles do you anticipate in adopting Artifact-based collaboration within your field, and what strategies could you employ to address them?
3. How does the concept of Artifacts as "extensions of human cognition" impact your view on the future of human-AI collaboration?

Action Items:

1. Experiment with using Artifacts for a complex writing or coding project
2. Develop a strategy for implementing Artifact-driven brainstorming in your next team meeting
3. Create an ethical guideline for Artifact use in your organization or personal work

Further Exploration:

- Research how other AI platforms are implementing persistent content features
- Explore potential integrations between Artifact-like systems and other productivity tools
- Search for innovative open-source projects that could further elevate the potential of AI Artifacts.

8

PRACTICAL APPLICATIONS IN VARIOUS FIELDS

8.1 Introduction: An Intersection

As we venture into the practical applications of AI collaboration, we find ourselves at the intersection of theory and real-world impact. This chapter will explore how the techniques and insights we've gained can be applied across diverse fields, transforming workflows, enhancing creativity, and pushing the boundaries of what's possible. We'll delve deep into each area, providing rich examples and practical strategies that you can adapt to your work.

8.2 Content Creation and Marketing

AI has become a key partner in the fast-changing realm of content creation and marketing, boosting human creativity and optimizing processes in extraordinary ways.

Collaborative Writing and Ideation

Imagine you're a marketing copywriter tasked with creating blog posts for a new eco-friendly product line. Here's a look at how you might leverage AI throughout the process:

Brainstorming with AI: Begin by using the divergent thinking technique we discussed in Chapter 5.

> **You:** "Generate 20 blog post ideas that appeal to environmentally conscious consumers interested in eco-friendly household products."
>
> **AI:** [Provides a list of 20 ideas]
>
> **You:** "Great, let's use the SCAMPER method on the top 3 ideas. Start with substituting key elements."

Audience Research: Leverage AI to uncover a more profound understanding of your audience.

> "Based on current trends, create three detailed personas of consumers interested in eco-friendly products. Include demographics, psychographics, and potential pain points."

Structuring Content: Once you have your ideas and audience insights, use the AI to help structure your content.

> "For the blog post '10 Simple Ways to Reduce Your Carbon Footprint at Home', create a detailed outline following this structure:
>
> - For each main point: One actionable tip (active voice, under ten words)
>
> - For each sub-point: Implementation cost ($ to $$$), time requirement (minutes/hours), and potential annual CO2 reduction (in metric tons)
>
> - For each example: One specific numerical case study or statistic from post-2020 research
>
> - For each tip: One common mistake to avoid.
>
> - Structure these as H2 headers for main points, H3 for sub-points, and bullet points for examples."

Drafting and Refining: Use the AI as a collaborative writing partner.

> **You:** "Based on this outline, write an engaging introduction for the blog post. Aim for a tone that's informative yet conversational."
>
> **AI:** [Provides a draft]
>
> **You:** "This is a good start, but let's make it more compelling. Rewrite the introduction to include a surprising statistic about carbon footprints and a personal anecdote that readers might relate to. Please supply a proper citation embedded in the article, so that I and others may reference it."

SEO Optimization: Leverage the AI's knowledge of SEO best practices.

> "Suggest 5-7 SEO-friendly keywords for this blog post about reducing carbon footprints at home. Then, propose how we can naturally incorporate these keywords throughout the article."

Creating Supplementary Content: Use the AI to generate ideas for accompanying visuals or social media posts.

> "Based on the content of this blog post, suggest ideas for an infographic that summarizes the key points visually. Then, draft five social media posts, each highlighting a different tip from the article, optimized for X's character limit."

Editing and Proofreading: Utilize AI for the final polish.

> "Review this draft for grammar, style, and clarity. Suggest improvements to make the language more engaging and accessible to our target audience."

Personalized Marketing Campaigns

AI can be instrumental in creating highly targeted, personalized marketing campaigns:

Customer Segmentation: Ask AI to analyze anonymized customer data and render detailed personas based on it.

> "Based on this customer purchase data and website interaction logs, identify 3-5 distinct customer segments. Create detailed personas for each, including demographics, preferences, behavioral patterns, and potential pain points."

Predictive Analytics: Leverage AI to forecast customer behavior and campaign performance.

> "Using the attached anonymized historical campaign data and current market trends, predict the performance of our upcoming email campaign for each customer segment. Provide best-case, worst-case, and most likely scenarios."

Tailored Messaging: Once you have your personas and predictions, use AI to craft tailored messages.

> "For the 'Eco-Conscious Millennial' persona, draft three email subject lines promoting our new bamboo toothbrush. Aim for an environmentally aware and slightly humorous tone. Then, create a short email body for each subject line, ensuring they align with the predicted interests of this segment."

A/B Testing: Employ AI to design, analyze, and iterate on A/B tests.

> **You:** "Design an A/B test for our product landing page, focusing on the call-to-action button. Suggest two variants to test against our control, explaining the psychological principle behind each variant."

> [After running the test]
>
> **You:** "Here are the results of our A/B test. Analyze these results, suggest three improvements for the underperforming versions, and recommend how we should iterate on this test."

Cross-Channel Campaign Coordination: Use AI to help coordinate messaging across various marketing channels.

> "Based on our campaign goals and target personas, suggest a coordinated marketing strategy that spans email, social media, and our blog. Include recommendations for timing, content themes, and creating a cohesive narrative across these channels."

Competitive Analysis: Leverage AI to gain insights into competitor strategies.

> "Analyze the attached social media content of our top three competitors over the past month. Identify common themes, engagement patterns, and unique approaches they use. Then, suggest three innovative ways we could differentiate our content strategy."

By leveraging AI in these ways, content creators and marketers can enhance their creativity, improve efficiency, and create more targeted, effective campaigns. The key is using AI as a collaborative tool that augments human creativity and strategic thinking rather than replacing these crucial skills.

8.3 Education and Personalized Learning

The field of education stands to benefit enormously from AI collaboration, offering opportunities for personalized learning experiences and enhanced educational content creation. Let's explore some practical applications in greater depth:

Curriculum Development and Lesson Planning

Course Outline Generation: Use AI to help structure course content.

> "Create a detailed outline for a 12-week introductory course on environmental science for high school students. Include main topics, subtopics, and suggested activities for each week. Ensure the course builds on previous knowledge and includes opportunities for practical application."

Diverse Learning Activities: Leverage AI to generate ideas for varied learning experiences.

> "For the topic 'Ecosystems,' suggest five different learning activities that cater to visual, auditory, and kinesthetic learners. For each activity, explain how it addresses different learning styles and what specific skills or knowledge it aims to develop."

Resource Curation: Use AI to find and evaluate educational resources. This is best done with a model that has internet access.

> "Find and summarize five recent, peer-reviewed articles on climate change suitable for a high school environmental science class. Provide a brief assessment of each article's complexity and relevance. Then, suggest how each article could be incorporated into a lesson plan, including potential discussion questions or activities."

Interdisciplinary Connections: Utilize AI to help create connections between different subjects.

> "Identify five ways that concepts from our environmental science course intersect with other high school subjects like biology, chemistry, economics, and social studies. Suggest a brief interdisciplinary activity for each connection to reinforce learning in both subjects."

Differentiated Instruction Planning: Use AI to help plan for diverse student needs.

> "For the attached lesson on photosynthesis, suggest modifications to accommodate students with different learning needs, including gifted students, students with dyslexia, and English language learners. Provide specific strategies and resources for each group."

Interactive Learning Tools

While AI-powered educational technologies offer innovative ways to enhance learning, adequate safeguards may be needed for using these approaches in practice.

Quiz Generation: Use AI to create engaging quizzes.

> "Generate a 10-question multiple-choice quiz on photosynthesis for 7th-grade students using Bloom's Taxonomy: 4 recall questions (knowledge level), three comprehension questions, two application questions, and one analysis question. Each question should have four options, include a clear explanation for the correct answer, and avoid negative phrasing. Align questions with Next Generation Science Standards (NGSS) MS-LS1-6."

Virtual Tutoring: Leverage AI for one-on-one tutoring sessions. This is best paired with a model that can accurately perform mathematical calculations like ChatGPT.

> "You are a patient and encouraging math tutor. Explain the concept of linear equations to a 9th-grade student struggling with the topic. Use simple language, provide step-by-step examples, and incorporate real-world applications. After each explanation, ask a checking question to ensure understanding."

Simulation Scenarios: Create interactive learning scenarios. Please note that adequate guardrails would be needed to use this approach in practice.

> "Design a text-based simulation of a historical event like the U.S. Declaration of Independence signing. Include decision points where students can explore different potential outcomes based on their choices. Provide historical context for each decision and explain the actual historical outcome after the student chooses."

Adaptive Storytelling for Language Learning: Use AI to create interactive stories that adjust to the learner's language level.

> "Create a short story in Spanish for a beginner language learner. Include decision points where the learner can choose different paths in the story. Adjust the language complexity based on the learner's choices, gradually introducing new vocabulary and grammar concepts."

Virtual Lab Experiments: Leverage AI to create virtual science experiments.

> "Design a virtual chemistry experiment to teach the concept of acid-base reactions. Describe the setup, procedure, and expected results. Include decision points where students can choose different chemicals or amounts and explain the outcomes of each choice. Suggest safety precautions that would be necessary in a real lab setting."

Educators can create more engaging, personalized, and compelling learning experiences by integrating AI into educational processes, catering to students' diverse needs, and enhancing educational outcomes. The key is to use AI to support and strengthen teachers' crucial role, not to replace the human elements of empathy, inspiration, and mentorship vital to education.

8.4 Business Productivity and Automation

AI collaboration can significantly enhance productivity, streamline processes, and drive innovation in the fast-paced business world. Let's explore some practical applications in greater depth:

Project Management and Planning

Project Breakdown: Use AI to help structure complex projects.

> "Break down the process of launching a new mobile app into main phases, tasks, and subtasks. Include estimated time frames for each major component and dependencies between tasks, and suggest key milestones. Also, identify potential bottlenecks and critical path elements."

Risk Assessment: Leverage AI to identify potential project risks.

> "Based on this project plan for launching a new product, identify potential risks and challenges. Categorize them by likelihood and potential impact, and suggest mitigation strategies for the top three risks. Also, propose a risk monitoring plan for the project's duration."

Resource Allocation: Use AI to optimize resource allocation.

> "Given these project requirements and this list of team member roles with their skills and availability, suggest an optimal allocation of human resources across the project timeline. Consider workload balancing, skill matching, and potential for skill development. Also, identify any skill gaps that need to be addressed through hiring or training."

Agile Sprint Planning: Utilize AI to assist with agile methodologies.

> "Based on our product backlog and team velocity, help plan our next two-week sprint. Suggest which user stories should be prioritized, how to break them down into tasks, and how to distribute them among team members. Also, propose some potential sprint goals."

Stakeholder Communication: Use AI to help craft effective project communications.

> "Draft a project status update for our executive stakeholders. Summarize our progress, highlight key achievements, address delays or issues, and outline the next steps. Ensure the language is concise and focuses on business impact."

Business Analytics and Reporting

Business Analytics and Reporting are potent tools for driving data-informed decision-making in organizations. While AI can significantly enhance these processes, it's crucial to approach mathematical analyses and interpretations cautiously.

Remember that models and algorithms, no matter how sophisticated, are based on assumptions and historical data that may not always reflect future realities.

As you explore the following AI-assisted analytics techniques, maintain a critical eye on the results, cross-verify findings when possible, and consider the context of your unique business environment. Ask AI to sift through anonymized customer data and build insightful, detailed profiles of your audience.

Data Analysis: Use AI to extract insights from complex data sets.

> "Analyze this sales data from the past year. Identify key trends, anomalies, and correlations. Suggest three actionable insights to improve sales performance. Also, identify external factors (like seasonal trends or economic indicators) that might influence our sales patterns."

Report Generation: Leverage AI to create comprehensive reports.

> "Based on this quarterly financial data, generate an executive summary highlighting key performance indicators, areas of concern, and recommendations for the next quarter. Include visualizations for the most important metrics and a brief competitive analysis based on publicly available data."

Predictive Analytics: Use AI for forecasting and scenario planning.

> "Using our historical sales data and current market trends, forecast our sales for the next six months. Provide best-case, worst-case, and most likely scenarios. Then, suggest how different marketing strategies might influence these forecasts."

Supply Chain Optimization: Use AI to improve supply chain efficiency.

> "Based on our current supply chain data, identify inefficiencies and bottlenecks. Suggest optimizations that could reduce costs or improve delivery times. Consider factors like demand forecasting, inventory management, and logistics routes."

Process Automation

Process Automation harnesses the power of AI to streamline operations, boost efficiency, and reduce human error across various business functions. Organizations can use their human resources for more critical and creative work by automating routine tasks and simplifying workflows.

This section explores how AI can enhance vital operational areas, from customer service to document processing.

As you delve into these AI-driven automation strategies, consider how they might be tailored to your business needs and integrated with existing systems to maximize their impact and drive sustainable improvements in productivity and quality.

Workflow Optimization: Use AI to analyze and improve business processes.

> "Review this customer service workflow. Identify bottlenecks and inefficiencies, then suggest three specific improvements to streamline the process. Estimate each suggestion's potential time and cost savings, and outline any challenges in implementing these changes."

Chatbot Development: Leverage AI to create intelligent chatbots.

> "Design a conversation flow for a customer service chatbot that can handle common queries about our product return policy. Include appropriate responses, escalation points for complex issues, and suggestions for personalizing interactions based on customer history."

Email Management: Use AI to categorize and prioritize emails.

> "Create rules for automatically categorizing and prioritizing incoming emails based on sender, subject line keywords, and content. Include suggestions for automated responses to common queries, and propose a system for flagging emails that might need urgent human attention."

Meeting Summarization: Utilize AI to capture key points from meetings.

> "Based on this transcript of our product development meeting, generate a concise summary highlighting key decisions, action items, and major discussion points. Delegate tasks to team members by the context of the conversation."

Intelligent Document Processing: Leverage AI for automated document analysis and data extraction.

> "Design a system to automatically extract key information from incoming invoices, regardless of format. Identify fields like invoice number, date, amount, and vendor details. Suggest how this system could integrate with our existing accounting software."

Digital Transformation Strategies

Digital Transformation Strategies are crucial for businesses aiming to stay competitive in today's rapidly evolving technological landscape. This section explores how AI can be leveraged to guide and accelerate your organization's digital evolution. AI offers powerful tools to inform decision-making and optimize your transformation journey, from assessing your current technology stack to planning change management and projecting ROI. The following examples demonstrate how AI can provide valuable insights and support at various stages of your digital transformation process.

Technology Stack Assessment: Use AI to evaluate and optimize your company's technology infrastructure.

> "Analyze the attached description of our current technology stack and compare it to industry best practices. Identify any outdated systems or potential security vulnerabilities. Suggest modernization strategies to improve efficiency, scalability, or data integration."

Change Management Planning: Leverage AI to plan for organizational changes during digital transformation.

> "Based on our planned system upgrade, create a change management strategy. Include a timeline for implementation, required training programs, potential resistance points, and communication strategies for different stakeholder groups."

ROI Projection: Use AI to forecast the return on investment for digital initiatives.

> "For our proposed CRM system upgrade, use Python to calculate the potential ROI over the next three years. Consider increased productivity, improved customer retention, and reduced maintenance costs. Provide best-case, worst-case, and most likely scenarios."

By integrating AI into various business processes, companies can enhance efficiency, make data-driven decisions, and better allocate human resources to more discretionary tasks. The key is to view AI as a powerful tool that augments human capabilities rather than a replacement for human judgment and creativity.

8.5 Research and Development

In research and development, AI collaboration can accelerate discovery, enhance analysis, and open new avenues of inquiry. Let's explore some practical applications in greater depth:

Literature Review and Synthesis

Comprehensive Search and Critical Evaluation: Utilize AI to conduct thorough literature searches and critically evaluate the results. While AI makes seemingly effortless work of churning through large batches of information, it's crucial to approach its output with discernment. This method is best paired with a model that has internet access.

> "Conduct a comprehensive search for peer-reviewed articles published in the last five years on CRISPR gene editing in cancer treatment. Please provide:
>
> A list of the 20 most influential papers based on citation count and journal impact factor
>
> For each paper: title, authors, publication date, journal name, and a summary of key findings
>
> Identification of emerging trends, methodological innovations, or conflicting results across these studies

Three specific areas where the current research seems to be converging

Three areas where significant disagreements or knowledge gaps persist."

> ***Note:*** *While AI can significantly accelerate the literature review process, always verify the existence and relevance of cited papers. Cross-reference AI-provided information with trusted academic databases and the original publications to ensure accuracy and comprehensiveness in your research.*

Summary and Analysis: Leverage AI to summarize and analyze research findings.

"For the attached paper on CRISPR gene editing in cancer treatment, provide a detailed summary, highlighting key findings, methodologies used, and potential implications for future research."

Gap Analysis: Use AI to identify research gaps.

"Based on this literature review, identify three key areas for further research in CRISPR applications in cancer treatment. Provide a rationale for each suggested research direction and outline potential study designs to address these gaps."

Cross-disciplinary Connections: Utilize AI to find relevant research from other fields.

You: "Identify research from fields outside of oncology that might have relevant applications to our work on CRISPR gene editing in cancer treatment. Look for analogous techniques or theoretical frameworks that could provide new perspectives on our research questions."

. . .

Experimental Design

Protocol Development: Use AI to assist in designing research protocols.

> "Help me design an experimental protocol to test the efficacy of a new drug compound on breast cancer cells. Include necessary controls, key variables to measure, and suggested timeframes. Also, propose methods to account for potential confounding factors and ensure reproducibility."

Statistical Analysis Planning: Leverage AI for planning data analysis.

> "For this drug efficacy experiment, suggest appropriate statistical tests to analyze the results. Explain why each test is suitable and what insights it can provide. Also, recommend sample sizes needed to achieve statistical significance, considering practical constraints."

Ethics Consideration: Use AI to identify potential ethical concerns.

> "Review this proposed clinical trial protocol. Identify any potential ethical issues and suggest modifications to address these concerns while maintaining the scientific integrity of the study. Consider informed consent, participant safety, data privacy, and potential long-term implications."

Simulation Design: Leverage AI to create research simulations.

> "Design a computational simulation to model the interaction of our proposed drug compound with its target protein. Include parameters for different physiological conditions and suggest how we could validate the simulation results against wet-lab experiments."

Data Interpretation and Visualization

Complex Data Analysis: Use AI to analyze complex datasets.

> "Analyze this anonymized dataset from 1000 cancer patients and 1000 healthy controls. Identify significant aspects, potential biomarkers, and any notable patterns or correlations. Suggest hypotheses explaining the observed patterns and propose follow-up analyses to test these hypotheses."

Visualization Suggestions: Leverage AI to suggest effective data visualizations.

> "Based on this anonymized data analysis, suggest five ways to represent the key findings visually. Include a brief explanation of each visualization's insights and how it could be used to communicate our results to scientific and lay audiences."

Interdisciplinary Connections: Use AI to identify potential cross-disciplinary applications.

> "Given these findings from our cancer genomics study, suggest three potential applications or implications for fields outside of oncology, such as evolutionary biology or drug development. For each suggestion, outline a potential collaborative research project exploring these connections."

Research Collaboration and Communication

Grant Proposal Development: Use AI to assist in crafting compelling grant proposals.

> "Based on our research plan, help me draft and review a compelling executive summary for a grant proposal. Highlight the innovation of our approach, its potential impact, and how it aligns with the funding agency's priorities. Suggest ways to make our proposal stand out from potential competitors."

Peer Review Preparation: Leverage AI to anticipate peer review questions and improve manuscripts.

> "Critically review our manuscript on CRISPR applications in cancer treatment. Identify potential weaknesses or areas that might attract criticism during peer review. Suggest ways to address these issues in the manuscript preemptively."

Conference Presentation Planning: Use AI to help structure and refine conference presentations.

> "Help me outline a 15-minute conference presentation on our recent findings. Suggest an engaging opening, key points to emphasize, and effective ways to present our complex data visually. Also, propose potential questions the audience might ask and how we could address them."

By integrating AI into the research process, scientists and R&D professionals can accelerate discovery, enhance the depth and breadth of their analyses, and potentially uncover insights that traditional methods might miss. However, it's crucial to remember that AI is a tool to augment human expertise and creativity, not replace it. The most groundbreaking discoveries will likely come from the insights between human intuition and AI's data-processing capabilities.

8.6 Healthcare and Medicine

Integrating AI in healthcare and medicine offers immense potential to revolutionize patient care, accelerate medical research, and enhance diagnostic accuracy. However, it's crucial to approach this integration with utmost caution and responsibility.

As we explore AI applications in this sensitive field, we must remain vigilant about the following:

- Ethical considerations
- Regulatory compliance
- Patient Privacy

Healthcare professionals and AI developers must remain ever mindful of potential biases in AI systems that could lead to disparities in care or misdiagnosis. Strict adherence to regulations such as HIPAA in the United States, GDPR in Europe, and other relevant local laws is non-negotiable.

Furthermore, while AI can be a powerful tool, it should always augment rather than replace the expertise and compassion of healthcare professionals.

As we delve into the following AI applications in healthcare, remember the delicate balance between innovation and the foundational principle of "first, do no harm."

Diagnostic Assistance

Image Analysis: Use AI to assist in interpreting medical imaging.

> "Help me develop an algorithm to analyze chest X-rays for signs of pneumonia. Include features to highlight areas of concern, quantify the severity of the condition, and compare the results to a database of known cases. Also, suggest how this tool could be integrated into radiologists' existing workflows."

Symptom Checking: Leverage AI for preliminary patient assessment.

> "Help me create a conversational AI system to conduct initial patient interviews, asking about symptoms and medical history. Design the system to flag potentially serious conditions that require immediate attention and provide a preliminary list of possible diagnoses for the healthcare provider to consider."

Predictive Diagnostics: Leverage artificial intelligence to forecast possible health concerns using patient information.

> "Create a framework to evaluate a patient's electronic health records, incorporating genetic data, lifestyle choices, and past medical history, to assess their likelihood of developing certain chronic diseases within the next 5 to 10 years. Additionally, provide tailored recommendations for preventive strategies based on each individual's risk assessment."

Medical Research

Clinical Trial Matching: Utilize artificial intelligence to connect patients with suitable clinical trials.

> "Help me design a system that can match cancer patients with suitable clinical trials based on their specific cancer type, genetic markers, treatment history, and other relevant factors. Include features to rank trials by potential benefit and proximity to the patient's location."

Drug Discovery: Leverage AI to accelerate the drug discovery process.

> "Create a model to screen a library of chemical compounds for potential effectiveness against a specific protein target implicated in Alzheimer's disease. Prioritize compounds based on predicted efficacy, safety profile, and ease of synthesis. Suggest a protocol for validating the top candidates in vitro."

Epidemiological Trend Analysis: Leverage AI to analyze disease trends.

> "Analyze this public health data from the past decade to identify trends in the spread of infectious diseases. Look for correlations between climate change, urbanization, and international travel patterns. Suggest areas where targeted interventions might be most effective."

Personalized Treatment Plans

Treatment Optimization: Use AI to tailor treatment plans to individual patients.

> "Help me design an AI system to analyze a cancer patient's genomic data, treatment history, and current health status to recommend personalized treatment options. Include a mechanism for incorporating the latest clinical research and drug approvals into the recommendation algorithm."

Medication Management: Leverage AI to improve medication adherence and prevent adverse interactions.

> "Create a smart medication management system to track a patient's medication schedule, potential drug interactions, and side effects. Include features for reminding patients to take their medications, alerting healthcare providers to potential issues, and suggesting dosage adjustments based on patient feedback and monitoring data."

Rehabilitation Planning: Use AI to create personalized rehabilitation programs.

> "Help me design an AI-powered system for creating personalized rehabilitation programs for stroke patients. The system should consider factors like the severity and location of the stroke, the patient's age and overall health, and their personal goals. Include features for tracking progress and adjusting the program based on the patient's recovery rate."

INTEGRATING AI into healthcare and medicine can improve diagnostic accuracy, treatment efficacy, and patient outcomes. However, it's crucial to approach this integration ethically, ensuring patient privacy, maintaining the human touch in healthcare, and using AI to support, not replace, healthcare professionals' expertise and judgment.

8.7 Environmental Science and Sustainability

Artificial intelligence can significantly contribute to tackling environmental issues and fostering sustainability. Here are some practical applications:

Climate Modeling and Prediction

Advanced Climate Models: Use AI to enhance climate prediction models.

> "Help me develop a machine learning model that can integrate data from various sources (satellite imagery, weather stations, ocean buoys) to improve short-term climate predictions. Focus on predicting extreme weather events with greater accuracy and longer lead times."

Impact Assessment: Leverage AI to model climate change impacts.

> "I need assistance developing a simulation to explore how a 2°C increase in global temperatures might affect agricultural productivity across various regions. The simulation should include factors like changing precipitation patterns, extreme weather events, and shifts in growing seasons."

Adaptation Strategy Planning: Use AI to develop climate adaptation strategies.

> "Based on climate change projections for the next 50 years, help me develop adaptation strategies for coastal cities. Consider factors such as sea-level rise, increased storm intensity, and potential changes in local ecosystems. Prioritize strategies based on cost-effectiveness and long-term resilience."

Resource Management

Smart Grid Optimization: Use AI to optimize energy distribution.

> "Help me design an AI system to manage a smart electricity grid. The system should balance supply and demand in real time, integrate renewable energy sources efficiently, and minimize energy waste. Include features for predicting peak demand periods and suggesting strategies to flatten the demand curve."

Water Conservation: Leverage AI for water resource management.

> "Help me develop an AI-driven system for urban water management. The system should predict water demand, detect leaks in the distribution system, and suggest conservation strategies. Include features for drought planning and water quality monitoring."

Sustainable Agriculture: Use AI to optimize farming practices.

> "Design an AI system that can analyze soil conditions, weather patterns, and crop data to provide personalized recommendations for farmers. Include features for optimizing irrigation, predicting pest outbreaks, and suggesting crop rotations to maintain soil health and maximize yield while minimizing environmental impact."

Biodiversity Conservation

Species Tracking: Use AI for wildlife monitoring and conservation.

> "Help me create an AI system that can identify and track endangered species using data from camera traps, satellite imagery, and citizen science reports. The system should be able to estimate population sizes, identify habitat threats, and suggest targeted conservation measures."

Ecosystem Modeling: Leverage AI to model complex ecosystems.

> "Help me develop a model of a coral reef ecosystem that can simulate the impacts of various stressors (ocean acidification, temperature rise, overfishing) on biodiversity and ecosystem health. Use the model to test the potential effectiveness of different conservation strategies."

Pollution Detection and Mitigation: Use AI to monitor and address pollution.

> "Design an AI system to analyze satellite imagery and sensor data to detect and track air and water pollution sources. Include features for predicting the spread of pollutants under different weather conditions and suggesting targeted mitigation strategies."

Utilizing AI in environmental science and sustainability allows us to gain deeper insights into intricate ecological systems, enhance prediction accuracy, and create more impactful conservation and resource management strategies. The essential approach is to blend AI technologies with specialized knowledge and ensure that the insights generated through AI translate into practical, actionable solutions.

8.8 Emerging Fields and Interdisciplinary Applications

As AI continues to evolve, it opens up new possibilities in emerging fields and creates opportunities for innovative interdisciplinary applications. Let's explore some of these cutting-edge areas:

Quantum Computing and AI

Quantum Machine Learning: Investigate the convergence of quantum computing and artificial intelligence.

> "Help me design a quantum machine learning algorithm that could potentially solve optimization problems in drug discovery more efficiently than classical algorithms. Explain the potential advantages and current limitations of this approach."

Quantum-Enhanced AI Security: Leverage quantum principles for AI security.

> "Outline a framework for using quantum key distribution to enhance the security of AI models, particularly in sensitive applications like financial forecasting or healthcare diagnostics. Discuss the potential benefits and implementation challenges."

Neurotechnology and Brain-Computer Interfaces

AI-Powered Neuroprosthetics: Use AI to enhance brain-computer interfaces.

> "Design an AI system that can interpret neural signals from a brain-computer interface to control a robotic prosthetic arm with high precision. Include features for adaptive learning to improve control over time and handle unexpected situations."

Cognitive Enhancement Applications: Explore AI applications in cognitive enhancement.

> "Suggest an AI-powered application that could improve human cognitive functions, like memory and attention. Be sure to consider the technical features and the ethical considerations surrounding this kind of technology."

AI in Space Exploration

Autonomous Space Probes: Design AI systems for deep space exploration.

> "Develop an AI system for an autonomous space probe designed to explore the moons of Jupiter. The system should be capable of making real-time decisions about navigation, data collection, and communication with Earth, considering the significant time delay in transmissions."

AI-Assisted Space Habitat Design: Use AI to optimize space habitat design.

> "Create an AI model to assist in designing a self-sustaining habitat for a Mars colony. The model should optimize for factors like resource utilization, energy efficiency, and psychological well-being of inhabitants under the constraints of the Martian environment."

AI in Nanotechnology

Nanobot Swarm Intelligence: Apply AI to control nanobot swarms.

> "Design an AI system to coordinate a swarm of medical nanobots for targeted drug delivery in the human body. The system should be able to navigate the bloodstream, locate specific cells or tissues, and coordinate the release of medication."

AI-Driven Nanomaterial Design: Use AI to discover new nanomaterials.

> "Develop an AI model that can predict the properties of novel nanomaterials based on their atomic structure. The model should suggest potential applications for these materials and guide experimental design for their synthesis."

AI in Synthetic Biology

Genome Design Optimization: Leverage AI for synthetic genome design.

> "Create an AI system that can optimize the design of synthetic genomes for specific functions, such as producing biofuels or pharmaceuticals. The system should consider factors like metabolic efficiency, genetic stability, and potential unintended consequences."

AI-Assisted Protein Engineering: Use AI to engineer novel proteins.

> "Design an AI model that can predict protein folding and engineer novel proteins with specific functions, such as enzymes for plastic degradation or proteins for targeted cancer therapy. Include a module for simulating the behavior of these proteins in various cellular environments."

By delving into these innovative and cross-disciplinary areas, we can expand the limits of AI, unlocking new possibilities in science, technology, and our understanding of the world. Achieving a harmony between innovation and ethical values is crucial, as we must ensure that our progress benefits humanity.

8.9 Ethical Considerations Across Fields

As we implement AI in different fields, it's important to reflect on the ethical considerations that come with these applications. Let's explore some key ethical considerations that span multiple fields:

Privacy and Data Protection

Healthcare Data Privacy: Ensure patient data confidentiality in AI-driven healthcare systems.

> "Design a framework for anonymizing patient data in AI-driven diagnostic systems. Include measures to prevent re-identification and ensure compliance with regulations like HIPAA. Also, suggest protocols for obtaining informed consent for AI analysis of patient data."

Consumer Privacy in Marketing: Balance personalization with privacy in AI-driven marketing.

> "Develop guidelines for using AI in personalized marketing that respect consumer privacy. Include recommendations for transparent data collection practices, opt-out mechanisms, and data retention and sharing limits."

Bias and Fairness

Fairness in AI-Driven Hiring: Ensure AI-assisted recruitment processes are fair and unbiased.

> "Develop a checklist for evaluating AI-based hiring systems to identify and address possible biases concerning gender, race, age, and other protected traits. Incorporate suggestions for utilizing diverse training datasets and conducting ongoing fairness evaluations."

Equitable Access to AI-Enhanced Education: Address potential disparities in AI-driven educational tools.

> "Propose strategies to ensure equitable access to AI-enhanced educational resources across different socioeconomic groups. Consider factors like internet access, device availability, and varying levels of digital literacy."

Transparency and Explainability

Explainable AI in Medical Diagnosis: Ensure AI-driven medical decisions are transparent and interpretable.

> "Develop a framework for explaining AI-driven medical diagnoses to healthcare professionals and patients. Include methods for visualizing the key factors influencing the AI's decision and suggest how to integrate these explanations into the clinical workflow."

Transparency in AI-Driven Financial Services: Ensure transparency in AI-powered financial decisions.

> "Create guidelines for explaining AI-driven decisions in financial services, such as loan approvals or insurance premium calculations. Include recommendations for providing clear, understandable explanations to customers and suggestions for regulatory compliance."

Accountability and Liability

AI Liability in Autonomous Vehicles: Address liability issues in AI-driven transportation.

> "Outline a framework for determining liability in autonomous vehicle accidents. Consider scenarios where AI systems make decisions leading to accidents and suggest approaches for balancing responsibility between manufacturers, software developers, and users."

AI Accountability in Environmental Management: Ensure accountability in AI-driven environmental decisions.

> "Please help me develop a framework for monitoring and assigning accountability for choices made by AI systems in environmental management, including aspects like resource distribution and pollution regulation. Include mechanisms for human oversight and intervention in critical decisions."

Human-AI Collaboration and Job Displacement

AI Integration in the Workplace: Address concerns about job displacement due to AI.

> "Develop strategies for integrating AI into workplace processes while minimizing job displacement. Include recommendations for reskilling programs, creating new AI-related job roles, and fostering a culture of human-AI collaboration."

Maintaining Human Expertise in AI-Assisted Fields: Ensure the preservation of human expertise in AI-augmented professions.

> "Propose guidelines for maintaining and developing human expertise in fields increasingly assisted by AI, such as radiology or legal research. Include strategies for continuous learning and skills development to complement AI capabilities."

Long-term Societal Impact

AI's Impact on Social Interaction: Address potential changes in human social behavior due to AI.

> "Analyze the potential long-term impacts of widespread AI personal assistants on human social interaction and emotional intelligence. Suggest strategies for maintaining healthy social skills in an AI-augmented world."

AI and Economic Inequality: Address potential exacerbation of economic disparities due to AI.

> "Examine how adopting AI technologies might impact economic inequality. Propose policy recommendations to ensure that the benefits of AI are distributed equitably across society."

8.10 Case Studies of Successful AI Integration

To illustrate the practical application of AI across various fields, let's examine some hypothetical case studies that demonstrate successful AI integration:

Healthcare: Personalized Treatment Planning

A major cancer research center implemented an AI-driven system for personalized treatment planning. The system recommended customized treatment plans by evaluating patient information, encompassing genetic details, medical history, and lifestyle choices.

Key Features:

- Collaboration with electronic health records to enable thorough analysis of patient information.
- Continuous learning from treatment outcomes to refine recommendations
- Explainable AI interface for oncologists to understand the reasoning behind suggestions

Results:

- 15% improvement in treatment efficacy
- 20% reduction in adverse side effects
- Increased patient satisfaction due to more personalized care

Ethical Considerations Addressed:

- Strict data privacy measures implemented
- Clear protocols for human oversight and final decision-making
- Regular audits for potential biases in treatment recommendations

Environmental Science: Smart Forest Management

A national forest service deployed an AI system for forest management and wildfire prevention.

Key Features:

- Analysis of satellite imagery, weather data, and ground sensor information
- Predictive modeling for fire risk and spread patterns
- Optimization of resource allocation for fire prevention and fighting

Results:

- 30% reduction in the area affected by wildfires
- More efficient use of limited resources for forest management
- Improved early warning system for nearby communities

Ethical Considerations Addressed:

- Transparency in decision-making processes for resource allocation
- Collaboration with Indigenous communities for traditional knowledge integration
- Regular environmental impact assessments of AI-driven interventions

Education: Adaptive Learning Platform

A leading education technology company developed an AI-powered adaptive learning platform for K-12 students.

Key Features:

- Customized learning journeys tailored to each student's performance.
- Real-time feedback and adjustments to teaching strategies
- Data-driven insights for teachers to identify struggling students

Results:

- 25% improvement in student test scores
- Increased student engagement and motivation
- More efficient use of teacher time for one-on-one support

Ethical Considerations Addressed:

- Strict data protection measures for student information
- Inclusive design to accommodate diverse learning needs
- Regular assessments to ensure the system doesn't reinforce or create new educational disparities

Business: AI-Driven Supply Chain Optimization

A global manufacturing company implemented an AI system for supply chain optimization.

Key Features:

- Real-time tracking and prediction of supply and demand fluctuations
- Automated inventory management and reordering
- Scenario planning for potential supply chain disruptions

Results:

- 20% reduction in inventory costs
- 30% improvement in on-time deliveries
- Enhanced resilience to supply chain shocks

Ethical Considerations Addressed:

- Fair treatment of suppliers, with transparent AI-driven decision-making
- Environmental impact considerations in logistics optimization
- Reskilling programs for employees affected by automation

These case studies showcase how AI can significantly enhance different sectors with care and ethical consideration. They emphasize the need to address moral issues, maintain human oversight, and prioritize real benefits for all parties involved.

8.11 Conclusion: The Art of AI Integration

AI is well-positioned to transform industries and solve previously unsolvable problems. The secret to effective AI integration is rooted in the technology and how we use it wisely, ethically, and harmoniously with human skills.

Remember the principles we've discussed throughout this book:

Visualize the AI's Attention: Always be mindful of what information the AI focuses on and how to direct its attention effectively.

Leverage the Push-Pull Dynamic: Use techniques to push critical information to the forefront of the AI's attention and pull relevant information when needed.

Iterative Refinement: Don't settle for the first output. Engage in collaborative refinement, guiding the AI to improve and expand upon its initial responses.

Cross-Pollination of Ideas: Look for opportunities to apply techniques from one field to another. For example, how AI assists in content creation might inspire new approaches in scientific research.

Ethical Considerations: Always consider the ethical implications of AI use, especially when dealing with sensitive information or making decisions that impact people's lives.

Human Insight is Key: Remember that AI optimizes human capabilities, not replaces them. The most powerful results come from combining AI capabilities with human creativity, intuition, and domain expertise.

Contextual Awareness: When applying AI, understand your field's specific context and constraints. What works in one domain may need to be adapted for another.

Continuous Learning: Stay updated on AI advancements and continuously experiment with new techniques and applications in your field.

Crystallize AI-Assisted Development: Leverage AI to enhance your software development process and create comprehensive test suites. This approach allows you to:

- Capture the combined wisdom of human expertise and AI insights
- Develop more reliable and robust software systems
- Reduce dependency on AI-invocations at runtime for every input
- Create a tangible, verifiable artifact of your AI collaboration
- Mitigate the risks of AI biases

By "crystallizing" AI-generated insights into well-structured code and tests, you create a more stable, predictable, and maintainable software ecosystem. This method bridges the gap between cutting-edge AI capabilities and traditional software engineering practices, ensuring that your AI-assisted developments stand the test of time and scrutiny.

As you apply these techniques in your work, you'll likely discover new and innovative ways to leverage AI collaboration. The examples in this chapter are just a glimpse of what you may achieve with AI. With practice and experimentation, you'll develop strategies for harnessing AI's power to enhance your work and push the boundaries of what's possible in your domain.

The future of human-AI collaboration is bright, and by mastering these techniques, you're positioning yourself at the forefront of this transformative wave. Remember, the goal isn't just to use AI as a tool but to develop a true partnership where humans and artificial intelligence complement and enhance each other for the good of humanity.

As we move forward into this AI-augmented future, let's do so with a commitment to responsible innovation, ethical considerations, and a focus on using these powerful technologies to address the most pressing challenges facing our world. The possibilities are limitless – it's up to us to shape them wisely and for the benefit of all.

CHAPTER 8 SUMMARY: PRACTICAL APPLICATIONS IN VARIOUS FIELDS

Key Concepts:

1. AI applications in content creation and marketing
2. AI integration in education and personalized learning
3. AI-driven business productivity and automation
4. AI in research and development
5. AI applications in healthcare and medicine

Reflection Questions:

1. How might you apply AI collaboration techniques in your work or study?
2. What hurdles do you anticipate in bringing AI into traditional sectors?
3. How do ethical considerations vary across different fields of AI application?

Action Items:

1. Identify one area in your work that could benefit from AI integration
2. Research case studies of AI implementation in your industry
3. Develop a small-scale pilot project to test AI collaboration in your field

Further Exploration:

- Explore industry-specific AI tools relevant to your field
- Research ethical guidelines for AI use in your professional domain
- Investigate emerging AI trends that could impact your industry in the next 5-10 years

9

MASTERING AI-DRIVEN PROJECTS: ORGANIZATION, ETHICS, AND IMPLEMENTATION

9.1 Introduction: Project Mindsets

AI has become a game-changer in the fast-paced world of technology, revolutionizing our methodologies in handling projects across diverse fields. This chapter focuses on harnessing the power of advanced language models like Claude 3.5 Sonnet or OpenAI's ChatGPT for project development and implementation. These AI collaborators represent a paradigm shift in how we conceive, execute, and refine our work.

Integrating AI and large language models (LLMs) into project workflows offers unprecedented creativity, efficiency, and problem-solving opportunities. It also brings distinct challenges that need careful thought and organized solutions. This chapter is a thorough guide for excelling in AI-driven projects, emphasizing the importance of effective collaboration with large language models (LLMs).

We'll explore the organizational systems needed to support AI collaboration, examine the ethical considerations underpinning our work, and delve into practical implementation strategies. By the end of this chapter, you'll have a robust framework for leveraging AI as a powerful ally in your projects while maintaining a human-centered and ethically sound approach.

9.2 Structured Organizational System for AI-Driven Projects

When embarking on projects involving AI collaboration, particularly with advanced language models, it's crucial to have a well-defined organizational system in place. This system should be flexible enough to accommodate AI's unique capabilities while providing the structure necessary for successful project completion.

Your organizational system should account for the following: - The AI capabilities discussed in Chapter 2 - The attention management principles from Chapter 3 - The communication strategies outlined in Chapter 4 - The ethical considerations detailed in Chapter 11

For field-specific organizational considerations, reference the relevant sections of Chapter 8.

Project Planning and Scoping

Define Clear Objectives: Start by clearly articulating your project's goals. What problem are you trying to solve? What outcomes do you hope to achieve? Having well-defined objectives will help guide your interactions with the AI model and ensure your efforts remain focused.

Identify AI-Suitable Tasks: Analyze your project to determine which tasks best suit AI collaboration. Language models excel at tasks such as:

- Research and information gathering
- Content generation and ideation
- Data analysis and pattern recognition
- Problem-solving and strategy development
- Code generation and debugging

Set Realistic Expectations: Understand the capabilities and limitations of the AI models you're working with. While they are mighty, they are not omniscient or infallible. Establish practical goals for what can be accomplished through collaboration with AI.

Create a Flexible Timeline: AI can often accelerate certain aspects of a project but may also introduce new complexities. Design a timeline that allows for iteration and refinement based on AI outputs.

Workflow Design

Iterative Approach: Design your workflow to be iterative, allowing for multiple rounds of AI interaction and human refinement. This approach maximizes the benefits of AI collaboration while ensuring human oversight and quality control.

Input Preparation: Develop a system for preparing high-quality inputs for the AI model. This might include:

- Crafting clear and specific prompts
- Gathering relevant background information
- Defining the desired format and style of outputs

Output Evaluation: Establish criteria and processes for evaluating AI-generated outputs. Consider factors such as:

- Relevance to project objectives
- Accuracy and factual correctness
- Coherence and logical flow
- Originality and creativity
- Ethical considerations

Refinement Loop: Implement a refinement loop where human expertise is used to guide and improve AI outputs. This might involve:

- Providing feedback to the AI model
- Editing and expanding on AI-generated content
- Fact-checking and verifying information
- Ensuring consistency with project goals and brand voice

Version Control: Implement a robust version control system such as Git to track changes and iterations in AI-generated content. This is crucial for maintaining a clear history of project development and facilitating collaboration.

Progress Tracking and Evaluation

Milestone Definition: Break your project into clear milestones that align with your objectives. These milestones should reflect both AI-assisted tasks and human-driven components of the project.

Regular Check-ins: If warranted, schedule regular check-ins to assess progress, address challenges, and adjust your approach as needed. These check-ins can be an opportunity to evaluate the effectiveness of your AI collaboration and make necessary adjustments.

Quantitative and Qualitative Metrics: Formally or informally develop quantitative and qualitative metrics to evaluate project success. These might include:

- Time saved through AI assistance
- Quality improvements in outputs
- Novel insights or solutions generated
- User or stakeholder feedback

Continuous Learning: Treat each project as a learning opportunity. Document insights about effective AI collaboration techniques, common pitfalls, and successful strategies. Use this knowledge to refine your approach in future projects.

Implementing a structured organizational system tailored to AI-driven projects creates a foundation for successful collaboration with language models. This system should emphasize flexibility, iteration, and a balance between AI capabilities and human expertise. In the next section, we'll delve deeper into how to use AI models within this organizational framework effectively.

9.3 Effective Use of AI Models

To maximize the benefits of collaborating with advanced language models like Claude 3.5 Sonnet or ChatGPT, you should first learn about their capabilities and how to use them effectively. This section will guide you through using AI models in your projects, from selecting a model to optimizing its performance in conversation.

Understanding AI Model Types

While this chapter focuses primarily on large language models (LLMs), it's worth briefly touching on the broader landscape of AI models:

Large Language Models (LLMs): These are the focus of our discussion and include models like Claude and ChatGPT. They excel at natural language processing tasks, including text generation, analysis, and problem-solving.

Computer Vision Models: Specialized in processing and analyzing visual information from images or videos.

Speech Recognition Models: Designed to convert spoken language into text.

Predictive Models: Used for forecasting future trends based on historical data.

Generative Models: Can create new content, including text, images, music or even video.

This chapter will concentrate on effectively using LLMs in project development.

Matching Models to Specific Tasks

Different language models have varying strengths and specializations. Here's how to match models to tasks:

General Knowledge and Reasoning: Models like Claude 3.5 Sonnet and GPT-4 excel at tasks requiring broad knowledge and complex reasoning. Use these for:

- Research synthesis
- Strategic planning
- Problem-solving
- Creative ideation

Code Generation and Analysis: While many LLMs can handle coding tasks, some are fine-tuned explicitly for this purpose. Consider using specialized coding models for:

- Writing and debugging code
- Explaining complex algorithms
- Optimizing existing code

Content Creation: Most advanced LLMs are well-suited for various content creation tasks, including:

- Writing articles, reports, or marketing copy
- Generating outlines and storyboards
- Crafting social media posts

Data Analysis and Interpretation: Large language models can help in examining and understanding intricate data sets by:

- Identifying patterns and trends
- Generating insights from data
- Crafting data-driven narratives

Language Translation and Localization: While specialized translation models exist, advanced LLMs can often handle translation tasks effectively, especially for context-sensitive content.

Optimizing Model Performance

To get the best results from your AI collaboration, consider these strategies:

Craft Clear and Specific Prompts: The quality of your input significantly affects the output. Write prompts that are:

- Specific about the task and desired outcome
- Rich in context and relevant information
- Clear about any constraints or requirements

Use System Messages Effectively: Many LLMs allow system messages that set the context for the entire conversation. Use these to:

- Define the AI's role or persona
- Establish critical parameters for the interaction
- Set overarching goals or constraints

Leverage Few-Shot Learning: Provide examples of the desired output format or style. This can significantly enhance the model's capacity to produce suitable responses.

Iterate and Refine: Don't expect perfect results on the first try. Use an iterative approach:

- Start with a broad prompt
- Analyze the output
- Refine your prompt based on the results
- Repeat until you achieve the desired outcome

Combine Model Strengths: For complex projects, consider using multiple models or instances of the same model for different aspects of the task. This allows you to leverage specific strengths for each component.

Temperature and Sampling Settings: If available, experiment with temperature and sampling settings:

- Lower temperature for more focused, deterministic outputs
- Higher temperature for more creative, diverse responses

Handling Model Limitations

As discussed in Chapter 2.6, LLMs have inherent limitations. Here are practical strategies to address these limitations in your projects:

Fact-Checking Implementation:

- Create a verification checklist for critical information
- Establish trusted external sources for validation
- Document and track which claims require verification
- Example: When using AI for medical content, maintain a database of peer-reviewed sources for verification

Managing Temporal Limitations:

- Create a system to track the AI's knowledge cutoff date
- Develop protocols for supplementing outdated information
- Build update workflows for time-sensitive content
- Example: For legal AI applications, maintain a separate database of recent law changes

Ensuring Consistency:

- Log and compare responses across similar queries
- Implement version control for AI-generated content
- Create templates for common tasks to ensure standardized outputs
- Example: For customer service applications, maintain a response database to check for consistency

Balancing Creativity and Accuracy:

- Define clear criteria for when to prioritize accuracy vs. creativity
- Establish different temperature settings for different types of tasks
- Create review protocols based on the task's requirements
- Example: Use lower temperature for technical documentation, higher for creative writing

Managing Uncertainty:

- Develop clear protocols for when AI should acknowledge uncertainty
- Create templates for expressing different levels of confidence
- Establish escalation paths for uncertain responses
- Example: Financial advice applications require explicit confidence levels for all predictions

Privacy Protection:

- Create data sanitization protocols
- Implement automated PII detection
- Establish clear guidelines for handling sensitive information
- Example: For healthcare applications, use anonymization techniques before processing patient data

EACH STRATEGY SHOULD BE TAILORED to your use case and integrated into your project's workflows. By implementing these practical approaches, you can effectively manage AI limitations while maximizing the benefits of AI collaboration.

9.4 Ethical Considerations in AI Model Selection and Use

Several key ethical considerations should guide your decision-making process when selecting and implementing AI models for your projects. This section focuses on practical ethical considerations specific to model selection and deployment.

Model Selection Criteria

Bias Assessment:

- Evaluate the model's training data sources and known biases
- Review available model cards or documentation for bias reports
- Test the model with diverse inputs to assess potential biases before deployment
- Consider whether the model's training data is appropriate for your use case

Transparency Level:

- Assess how well the model can explain its outputs
- Determine if the model's level of transparency meets your project's requirements
- Consider whether more interpretable models would be better suited for your use case
- Match the model's explainability features to stakeholder needs

Privacy Features:

- Evaluate the model's data-handling practices
- Check for built-in privacy protection features
- Assess whether the model retains or learns from user inputs
- Consider whether hosted or local deployment better serves your privacy needs

Implementation Safeguards

Output Monitoring:

- Implement systematic checks for biased or inappropriate outputs
- Set up regular audits of model performance across different user groups
- Create feedback loops to identify and address emerging issues
- Establish clear thresholds for human review

Access Controls:

- Define who can access and use the model
- Set up appropriate authentication and authorization systems
- Create audit trails for model usage
- Implement role-based access for different levels of model functionality

Error Prevention:

- Design input validation systems to prevent misuse
- Implement output filtering for sensitive content
- Create fallback mechanisms for when the model fails
- Set up alerts for unusual or potentially problematic model behavior

Deployment Guidelines

Integration Checks:

- Verify that model integration maintains ethical safeguards
- Test the complete system for unintended consequences
- Ensure error handling preserves user privacy and security
- Validate that monitoring systems work as intended

User Communication:

- Indicate when users are interacting with AI
- Provide appropriate disclaimers about model limitations
- Explain how user data will be handled
- Offer channels for user feedback and concerns

Maintenance Plan:

- Schedule regular ethical impact assessments
- Plan for model updates and version control
- Establish procedures for addressing discovered issues
- Create documentation requirements for system changes

Tip: *See Chapter 11 for a comprehensive framework on ethical AI implementation and broader ethical considerations. This chapter focuses on practical considerations during model selection and deployment.*

9.5 Recap of Communication Strategies for AI Collaboration

Effective communication is crucial when collaborating with AI models on projects. This involves not only how we communicate with the AI but also how we discuss AI-driven processes with team members and stakeholders. Let's recap strategies for clear, productive communication in AI-driven projects.

Crafting Effective Prompts

How we communicate with AI models through prompts significantly influences the quality and relevance of their outputs. Here are strategies for crafting effective prompts:

Be Specific and Clear:

- Clearly state the task or question you want the AI to address.
- Provide necessary context and background information.
- Declare the desired response structure (e.g., a list).

"Examine how remote work influences employee productivity. Consider both positive and negative effects. Structure your response in bullet points, covering at least five key points. Provide a brief explanation and a potential data point or statistic supporting each point."

Use Step-by-Step Instructions:

- Divide intricate tasks into smaller, more achievable steps.
- Guide the AI through your thought process.

"Let's develop a marketing strategy for a new electric car charging station. Follow these steps:

1. Identify the target audience

2. List three key product features and their benefits

3. Suggest three marketing channels and explain why they're suitable

4. Propose a tagline that emphasizes the eco-friendly aspect

5. Outline a simple 30-day launch plan."

Leverage Chain-of-Thought Prompting:

- Encourage the AI to show its reasoning process.
- This can lead to more accurate and thoughtful responses.

"Solve this logic puzzle: 'If all Zorks are Flurbs, and no Flurbs are Pixies, can a Zork be a Pixie?' Please explain your thought process step by step before providing the final answer."

Use Examples (Few-Shot Learning):

- Provide examples of the type of response you're looking for.
- This can help the AI understand the desired format and style.

"Write a product description for a new fitness tracker. Here's an example of the style we're aiming for:

'The PeakPulse Elite: Your companion for peak performance. With an aerospace-grade titanium case, advanced sleep monitoring, and a vibrant, always-on display, the Elite doesn't just track your workouts —it transforms them. Experience real-time coaching, precise heart rate analytics with our medical-grade sensors, and confidence knowing your device is as reliable on mountain trails as in the gym.'

Write a similar description for our new smart running shoes, the FlexStride Pro."

Explaining AI Concepts to Non-Technical Stakeholders

When working on AI-driven projects , you'll often need to explain complex AI concepts to team members or stakeholders who may need a technical background. Here are some strategies:

Use Analogies and Metaphors:

- Compare AI concepts to familiar ideas or processes.
- Example: "Working with an AI language model is like having a knowledgeable research assistant. It can help gather and synthesize information quickly, but you must still guide its work and verify its outputs."

Focus on Capabilities and Limitations:

- Communicate what the AI can and cannot do.
- Avoid technical jargon and focus on practical implications.
- Example: "Our AI can quickly generate draft marketing copy based on key points we provide. However, it doesn't have real-time knowledge of current events or our specific brand voice, so human review and editing are crucial."

Provide Concrete Examples:

- Show real examples of AI inputs and outputs.
- Demonstrate how human expertise complements AI capabilities.
- Example: Walk through a simple task, showing the prompt given to the AI, its response, and how a human would refine or use that output.

Address Common Misconceptions:

- Proactively clarify common misunderstandings about AI.
- Example: "While our AI model is very advanced, it doesn't 'think' or 'understand' as humans do. It's a sophisticated pattern recognition tool that can generate human-like text, but it doesn't have consciousness or true comprehension."

Emphasize the Human Role:

- Highlight the importance of human oversight and decision-making.
- Example: "The AI is a powerful tool that can enhance our productivity and creativity, but it's not a replacement for human expertise. We use it to augment our capabilities, not to make decisions for us."

Documenting AI Processes and Decisions

Thorough documentation is crucial for transparency, reproducibility, and continuous improvement in AI-driven projects. Here's what to focus on:

Prompt Library:

- Maintain a catalog of effective prompts for various tasks.
- Include notes on when and how to use each prompt.

Decision Log:

- Record key decisions made based on AI outputs.
- Document the reasoning behind accepting, modifying, or rejecting AI suggestions.

Version Control:

- Keep track of different versions of AI-generated content using tools like Git.
- Document the refinement process from the initial AI output to the final product.

Ethics and Bias Checks:

- Document regular checks for potential biases or ethical issues in AI outputs.
- Record any mitigation strategies implemented.

Performance Metrics:

- Track essential metrics to assess the effectiveness of AI collaboration.
- This could include saving time, improving quality, or generating novel insights.

Lessons Learned:

- After each project, document insights about effective AI collaboration techniques.
- Note any challenges encountered and how they were addressed.

Facilitating Human-AI Interaction

Creating a smooth workflow between human team members and AI models is key to successful projects. Consider these strategies:

Establish Clear Workflows:

- Define when and how AI should be used in your processes.
- Create guidelines for human review and refinement of AI outputs.

Encourage Experimentation:

- Encourage a culture where team members feel at ease exploring and experimenting with AI technologies.
- Share successful prompts and techniques across the team.

Provide Training:

- Offer training sessions on effective, prompt engineering and AI collaboration.
- Keep the team updated on new AI capabilities and best practices.

Feedback Loops:

- Establish mechanisms for team members to share their insights on AI performance.
- Use this feedback to refine prompts and processes continually.

Collaborative Platforms:

- Consider using platforms that facilitate seamless integration of AI into workflows.
- Ensure these platforms allow for easy sharing and iteration of AI-generated content.

UTILIZING these communication strategies can foster a more efficient and transparent atmosphere for collaborating with AI. This method guarantees that all participants grasp the significance of AI in your projects and can play a role in its successful implementation. The upcoming section will cover crucial data security and privacy aspects regarding AI collaboration.

9.6 Data Security and Privacy in AI Collaboration

Prioritizing data security and privacy is essential in project development involving advanced AI models. This section will examine crucial factors and practical approaches to safeguarding the confidentiality and integrity of your data during AI collaborations.

Understanding the Landscape

Before diving into specific practices, it's essential to understand the broader context of data security and privacy in AI collaboration:

Model Training Data: Large language models are trained on vast public data. While they don't retain specific information from this training data, they may inadvertently reproduce sensitive or copyrighted information.

Input Data Privacy: Any information you input as a prompt becomes part of the interaction and may be used to improve the model (depending on the service's policies).

Output Data Ownership: The ownership and copyright of content created by AI can be complicated and often depend on the specific terms of use set by the AI service.

Regulatory Compliance: Depending on your industry and geographical area, you may be required to adhere to particular data protection laws, such as GDPR, CCPA, or HIPAA.

Data Protection Regulations and Compliance

Ensuring adherence to applicable data protection laws is essential when incorporating AI into your projects:

Identify Applicable Regulations:

- Assess the applicable data protection laws for your project by considering your location, industry, and how you manage data.
- Common regulations include GDPR (European Union), CCPA (California), HIPAA (U.S. healthcare), and industry-specific standards.

Data Classification:

- Classify your data based on sensitivity (e.g., public, internal, confidential, highly confidential).
- Establish clear guidelines on what data types can be used with AI models.

Data Minimization:

- Only use the minimum amount of data necessary for the task.
- Avoid inputting full datasets or large amounts of sensitive information into AI models.

Consent and Transparency:

- If using customer or user data, ensure you have proper consent for AI processing.
- Be transparent about how AI is used in your processes, especially when it affects individuals.

Right to Explanation:

- For decisions significantly impacting individuals, be prepared to explain how AI contributed to those decisions.
- Maintain clear documentation of AI inputs, outputs, and human oversight.

Data Retention and Deletion:

- Understand the data retention policies of the AI services you're using.
- Implement processes for secure deletion of data when no longer needed.

Secure Data Handling Practices

Implementing robust security measures is essential when collaborating with AI:

Data Encryption:

- Ensure secure, encrypted connections when interacting with AI services.
- Confirm that robust encryption is employed for both transmitted and stored data.

Access Control:

- Implement strict access controls for AI tools and any sensitive data.
- Use multi-factor authentication to access AI platforms.
- Regularly audit and update access permissions.

Secure Development Practices:

- When creating applications that engage with AI models, adhering to secure coding practices is essential.
- Regularly update and patch any software or libraries used in your AI workflow.

Data Masking and Anonymization:

- Data masking techniques obscure sensitive information when working with AI models.
- Anonymize data where possible, especially when using real datasets for testing or fine-tuning.

Secure Storage:

- Store any sensitive data or AI outputs in secure, access-controlled environments.
- Use reputable cloud services with strong security measures if storing data off-premises.

Incident Response Plan:

- Create and update an incident response strategy tailored for data breaches involving AI.
- Evaluate and refresh this plan regularly to maintain its effectiveness.

Privacy-Preserving AI Techniques

Consider implementing privacy-preserving techniques in your AI collaboration:

Federated Learning:

- This technique allows training AI models on distributed datasets without centralizing the data.
- It can be beneficial for projects involving sensitive or regulated data.

Differential Privacy:

- Implement differential privacy techniques to add controlled noise to data or model outputs.
- This can help prevent the extraction of individual data points from aggregated results.

Homomorphic Encryption:

- This advanced technique allows for computations on encrypted data without decrypting it.
- While computationally intensive, it can be valuable for applications that process confidential information.

Secure Multi-Party Computation:

- This enables several participants to collaboratively evaluate a function based on their inputs while ensuring confidentiality.
- It can be helpful for collaborative projects involving sensitive data from multiple sources.

Ethical Data Collection and Usage

Ensuring ethical data practices is crucial for maintaining trust and compliance:

Data Source Verification:

- Verify the legitimacy and ethical standing of your data sources.
- Ensure you have the right to use the data for your intended purposes.

Bias Detection and Mitigation:

- Regularly assess your data and AI outputs for potential biases.
- Implement strategies to mitigate discovered biases, such as data augmentation or model fine-tuning.

Transparency in Data Usage:

- Communicate how data will be used in AI processes.
- Provide options for individuals to opt out of using their data for AI training or processing where appropriate.

Ethical Review Process:

- Implement an ethical review process for AI projects, especially those using sensitive data or making impactful decisions.
- Consider forming an ethics board for large-scale or sensitive projects.

Continuous Monitoring:

- Regularly audit your AI processes for ethical considerations.
- Stay informed about evolving ethical standards and best practices in AI.

Best Practices for AI Collaboration Platforms

When choosing and using AI collaboration platforms:

Vendor Assessment:

- Thoroughly assess the security and privacy practices of AI service providers.
- Review their data handling policies, encryption methods, and compliance certifications.

Data Residency:

- Be aware of where your data is processed and stored using cloud-based AI services.
- Ensure compliance with any data residency requirements applicable to your project.

API Security:

- Implement proper API security measures using APIs to interact with AI models.
- Use API keys securely and regularly rotate them.

Audit Logging:

- Implement comprehensive logging of all AI interactions.
- Consistently monitor logs to identify any unusual activities or possible data breaches.

Confidentiality Agreements:

- Consider implementing confidentiality agreements or NDAs when working with external AI services on sensitive projects.

FOCUSING on data security and privacy in your AI partnerships can improve adherence to regulatory standards, foster stakeholder trust, and help to safeguard your organization's essential information assets.

9.7 Conclusion: Foundations for Success

As we conclude this chapter on the organizational and planning aspects of AI-driven projects, it's clear that success lies in the careful balance of structure and flexibility. The frameworks and strategies we've explored provide a solid foundation for integrating AI into your project workflows while maintaining ethical standards and ensuring data security.

Key takeaways include:

- The importance of establishing transparent organizational systems that can adapt to AI's unique capabilities
- The critical role of ethical considerations in every aspect of AI project planning
- The necessity of robust communication strategies for effective AI collaboration
- Data security and privacy in AI-driven projects are paramount

Remember that these foundations are not rigid structures but adaptable frameworks that should evolve with your projects and the rapidly advancing AI landscape. The key is to balance systematic approaches and the flexibility needed to leverage AI's innovative potential.

As you progress with your AI-driven projects, use these frameworks as starting points, adapting and refining them based on your specific needs and experiences. The most successful AI implementations will come from those who can effectively blend structured organization with creative problem-solving.

In the next chapter, we'll explore how to stay current with AI developments and grow your AI capabilities, building upon the solid foundation we've established here.

CHAPTER 9 SUMMARY: MASTERING AI-DRIVEN PROJECTS: ORGANIZATION AND PLANNING

Key Concepts:

1. Structured organizational systems for AI project management
2. Effective techniques for AI model selection and use
3. Ethical considerations in AI project development
4. Communication strategies for AI collaboration
5. Data security and privacy frameworks

Reflection Questions:

1. How might you adapt your current project management practices to incorporate AI collaboration?
2. What challenges do you anticipate in maintaining ethical standards while driving innovation?
3. How does the need for data security impact your approach to AI implementation?

Action Items:

1. Develop an AI-specific project management template for your team
2. Create a communication protocol for AI collaboration
3. Establish data security and privacy guidelines for AI projects

Further Exploration:

- Research project management tools with AI integration capabilities
- Explore frameworks for ethical AI development
- Investigate advanced data protection techniques for AI projects

10

STAYING CURRENT AND GROWING WITH AI

10.1 Introduction: Evolution and Growth

Staying current and effectively growing your AI capabilities has become crucial for long-term success. This chapter focuses on AI implementation: how to stay informed about new developments, integrate AI into existing workflows, and ensure your organization is ready for increasingly sophisticated AI applications.

The rapid advancement of AI technologies presents both opportunities and challenges. While new capabilities emerge regularly, offering powerful tools for innovation and efficiency, keeping pace with these changes requires deliberate strategy and continuous learning. This chapter will provide practical frameworks for managing this evolution and growing your AI capabilities sustainably.

We'll explore strategies for staying informed about AI developments, examine methods for seamless integration into existing workflows, and provide comprehensive tools for assessing and improving your AI readiness. By the end of this chapter, you'll have a clear roadmap for growing your AI capabilities while maintaining effectiveness and ethical standards in your implementations.

10.2 Staying Informed About AI Developments

Artificial intelligence, particularly in areas like large language models, is evolving at a breakneck pace. Keeping up with the most recent advancements is essential for effectively utilizing AI in your projects and staying ahead of the competition. This section will explore strategies for keeping up-to-date with AI advancements and integrating new knowledge into your work.

Key Resources for AI News and Research

Academic Journals and Preprint Servers:

- arXiv (arxiv.org): A popular preprint server where researchers often post their latest AI papers.
- Journal of Artificial Intelligence Research (JAIR)
- IEEE Transactions on Pattern Analysis and Machine Intelligence

AI-Focused News Websites:

- MIT Technology Review's AI section
- AI News (artificialintelligence-news.com)
- The AI Journal (aijourn.com)

Tech News Sites with Strong AI Coverage:

- Wired
- TechCrunch
- The Verge

AI Research Lab Blogs and Publications:

- OpenAI Blog
- Google AI Blog
- DeepMind Blog
- Microsoft Research AI Blog
- Anthropic Newsroom

AI Podcasts:

- Lex Fridman Podcast
- The AI Podcast by NVIDIA
- The TWIML AI Podcast

Social Media and Professional Networks:

- X (formerly Twitter): Follow prominent AI researchers and organizations
- LinkedIn: Join AI-focused groups and follow thought leaders
- GitHub: Track popular AI repositories and projects

AI Newsletters:

- Import AI by Jack Clark
- The Algorithm by MIT Technology Review
- AI Weekly

Participating in AI Communities and Conferences

Engaging with the AI community can provide valuable insights and networking opportunities:

Online Communities:

- Reddit: r/MachineLearning, r/artificial
- Stack Overflow: AI and Machine Learning tags
- Kaggle: Participate in AI competitions and discussions

Conferences and Workshops:

- NeurIPS (Neural Information Processing Systems)
- ICML (International Conference on Machine Learning)
- AAAI Conference on Artificial Intelligence
- ACL (Association for Computational Linguistics) for NLP-focused events

Local Meetups and User Groups:

- Look for AI, Machine Learning, or Data Science meetups in your area
- Consider starting your own if none exist

Hackathons and AI Challenges:

- Participate in AI-focused hackathons to apply your skills and learn from others
- Engage in online AI challenges on platforms like Kaggle or DrivenData

Open Source Contributions:

- Contribute to open-source AI projects on GitHub
- This can provide hands-on experience with cutting-edge AI tools and techniques

Continuous Learning Strategies for AI Professionals

Continuously learn more about AI advancements:

Online Courses and MOOCs:

- Coursera, edX, and Udacity offer advanced AI and ML courses
- Fast.ai provides practical deep-learning courses
- Keep an eye out for courses on emerging AI topics

Technical Workshops and Webinars:

- Attend workshops offered by AI tool vendors (e.g., NVIDIA, Google Cloud)
- Participate in webinars hosted by AI research organizations

Reading Research Papers:

- Develop a habit of regularly reading AI research papers
- Start with survey papers to get an overview of a field, then dive into specific topics

Implement and Experiment:

- Try implementing new AI techniques in small projects or experiments
- Use platforms like Google Colab or Jupyter Notebooks for quick prototyping

Cross-disciplinary Learning:

- Explore related fields like cognitive science, neuroscience, or philosophy of mind
- Gaining knowledge in these areas can offer important perspectives on the development of AI

Teach and Share Knowledge:

- Write blog posts or articles about your AI experiences and learnings
- Give presentations at local meetups or conferences
- Teaching others is an excellent way to solidify your understanding

Set up AI News Alerts:

- Set up Google Alerts or comparable services to receive updates on particular AI topics of interest.
- This can help you stay on top of breaking news in your areas of interest

Evaluating and Adopting New AI Technologies

With the rapid pace of AI development, it's crucial to have a systematic approach to evaluating and potentially adopting new technologies:

Establish an AI Technology Radar:

- Create a system for tracking emerging AI technologies relevant to your field
- Categorize technologies based on their potential impact and readiness for adoption

Proof of Concept (PoC) Process:

- Develop a streamlined process for running PoCs on promising new AI technologies
- Set clear criteria for what constitutes a successful PoC

Benchmark Against Existing Solutions:

- When evaluating new AI technologies, benchmark them against your current solutions
- Consider factors like performance, cost, ease of integration, and maintenance requirements

Consider Ethical Implications:

- Assess the ethical implications of new AI technologies before adoption
- Ensure they align with your organization's values and ethical guidelines

Scalability and Future-proofing:

- Evaluate how well new AI technologies can scale with your needs
- Consider their potential for future development and support

Community and Ecosystem:

- Look at the community and ecosystem around new AI technologies
- A strong, active community can be invaluable for support and ongoing development

Integration with Existing Workflows:

- Assess how easily new AI technologies can be integrated into your existing workflows
- Consider the training and adaptation time required for your team

By staying informed about AI developments and systematically evaluating new technologies, you can ensure that your AI-driven projects remain at the cutting edge. This continuous learning and adaptation process is critical to long-term success in the rapidly evolving field of AI.

10.3 Integrating AI into Existing Workflows

Integrating AI, particularly advanced language models like Claude or ChatGPT, into your existing workflows is crucial for maximizing the benefits of AI collaboration. This section will guide you through seamlessly incorporating AI into your projects.

Assessing Current Workflows

Before integrating AI, it's important to understand your current processes thoroughly:

Process Mapping:

- Create detailed maps of your existing workflows
- Identify key stages, decision points, and bottlenecks

Task Analysis:

- Break down workflows into specific tasks
- Categorize tasks based on their complexity, frequency, and importance

Identify Pain Points:

- Pinpoint areas where current processes are inefficient or prone to errors
- Identify tasks that are repetitive and require considerable time and resources.

Data Flow Analysis:

- Map out how information flows through your current processes
- Identify where data is input, processed, and output

Stakeholder Input:

- Gather feedback from team members involved in current workflows
- Understand their challenges and ideas for improvement

Identifying AI Integration Opportunities

Once you have a clear picture of your current workflows, you can start identifying where AI can add value:

Task Suitability Assessment:

- Evaluate which tasks are well-suited for AI assistance
- Consider tasks that involve:
  - Natural language processing (e.g., content creation, summarization)
  - Pattern recognition and analysis
  - Repetitive or time-consuming processes
  - Complex decision-making requiring large amounts of data

Quick Wins:

- Start with tasks where AI can provide immediate, tangible benefits
- Look for areas where AI can significantly reduce time or improve accuracy reliably

Strategic Long-term Opportunities:

- Identify areas where AI could enable new capabilities or business models
- Consider how AI could help scale operations or enter new markets

Risk Assessment:

- Evaluate the potential risks of integrating AI into different parts of your workflow
- Consider factors like data sensitivity, impact of errors, and regulatory compliance

Resource Evaluation:

- Assess the resources (time, budget, expertise) required for AI integration
- Prioritize opportunities based on potential ROI and resource availability

Designing AI-Enhanced Workflows

Once you've identified integration opportunities, it's time to design your new AI-enhanced workflows:

Define Clear Objectives:

- Set specific goals for each AI integration (e.g., reduce processing time by 50%, improve accuracy by 20%)
- Ensure these goals align with broader project or organizational objectives

Human-AI Collaboration Model:

- Create applications and processes to route work to both humans and AIs efficiently
- Clearly define roles and handoff points between AI and human team members

AI Prompt Design:

- Develop standardized prompts for common tasks
- Create guidelines for effective, prompt engineering

Output Validation Process:

- Implement processes for validating and refining AI outputs
- Define criteria for when human review is necessary

Feedback Loops:

- Design mechanisms for continually improving AI performance based on human feedback
- Implement systems for tracking and analyzing AI performance over time

Error Handling and Escalation:

- Develop protocols for handling AI errors or unexpected outputs
- Create clear escalation paths for complex issues

Version Control and Documentation:

- Implement version control for AI-generated content
- Maintain comprehensive documentation of AI-assisted processes

Implementation Strategies

Implementing AI into your workflows requires a thoughtful approach:

Phased Rollout:

- Begin with a pilot project or a limited-scale rollout.
- Gradually expand AI integration based on initial results and learnings

Training and Onboarding:

- Provide comprehensive training for team members on working with AI
- Create user-friendly guides and resources for AI collaboration

Continuous Monitoring:

- Implement systems for monitoring AI performance and integration effectiveness
- Regularly review metrics and gather feedback from users

Iterative Improvement:

- Continuously refine and optimize AI-enhanced workflows
- Be open to adjusting your approach based on the results you observe.

Technical Integration:

- Work with IT teams to ensure smooth technical integration of AI tools
- Address any compatibility issues with existing systems

Change Management:

- Develop a change management strategy to help team members adapt to new AI-enhanced workflows
- Communicate the benefits and address concerns proactively

Overcoming Common Challenges

Be prepared to address these common challenges when integrating AI into workflows:

Resistance to Change:

- Address fears about job displacement
- Highlight the importance of AI in enhancing human roles rather than laying them off. Stand by your word on this, as actions speak louder than words.

Data Quality and Availability:

- Ensure you access reliable and pertinent data for effective AI training and functioning.
- Implement data cleaning and preparation processes

Integration Complexity:

- Anticipate the technical hurdles involved in merging AI with existing legacy systems.
- Consider using API-based solutions for easier integration

Ethical Concerns:

- Handle ethical issues promptly, including bias and privacy concerns.
- Implement ethical guidelines and review processes

Expectations Management:

- Set realistic expectations about AI capabilities and limitations
- Educate stakeholders on what AI can and cannot do

Skill Gaps:

- Recognize and tackle any skill deficiencies within your team regarding AI proficiency.
- Consider hiring AI specialists or upskilling existing team members

Cost Management:

- Carefully manage costs associated with AI integration, including licensing, computing resources, and potential productivity dips during the transition

By carefully incorporating AI into your current workflows, you can significantly improve your project results and productivity. Keep in mind that achieving successful integration is a continuous journey that necessitates regular monitoring, learning, and adapting. The following section will discuss evaluating your organization's preparedness for adopting AI.

10.4 AI Readiness Checklist for Projects

Assessing your readiness is crucial before fully committing to AI-driven project development. This checklist will help you evaluate your preparedness and identify areas needing attention before embarking on AI-intensive projects.

Technical Infrastructure Assessment

Computing Resources:

□ Do you have access to sufficient computing power for AI tasks?

□ Is cloud computing an option for scalability?

Data Storage and Management:

□ Do you have adequate data storage systems in place?

□ Are your data management practices compliant with relevant regulations?

AI Development Environment:

□ Do you have the necessary software and tools for AI development

□ Are version control systems in place for both code and data?

API and Integration Capabilities:

□ Can your existing systems integrate with AI tools and services?

□ Do you have experience working with APIs for AI services?

Security Infrastructure:

□ Are robust cybersecurity measures in place to protect AI systems and data?

□ Do you have protocols for secure AI model deployment?

Data Readiness Evaluation

Data Availability:

□ Can you access sufficient, relevant data for your AI projects?

□ Do any gaps in your data need to be addressed?

Data Quality:

□ Is your data clean, consistent, and accurate?

□ Do you have processes in place for data cleaning and preparation?

Data Governance:

□ Have clear data governance policies been established?

□ Do you have systems for ensuring data privacy and compliance?

Data Labeling:

□ If needed, do you have resources for data labeling?

□ Are there processes in place to ensure label quality and consistency?

Data Pipeline:

□ Is there a robust data pipeline for collecting and processing new data?

□ Can your data pipeline handle the volume and velocity of data required for AI projects?

Skill Gap Analysis

AI Expertise:

☐ Do you have team members with AI and machine learning expertise? [ ] Is there a plan for ongoing AI education and training?

Domain Knowledge:

☐ Do your AI practitioners have sufficient domain knowledge?

☐ Is there a system for knowledge transfer between domain experts and AI practitioners?

Prompt Engineering Skills:

☐ Are team members trained in practical prompt engineering for language models?

☐ Is there a process for sharing and improving prompt engineering techniques?

AI Ethics Understanding:

☐ Is there awareness of AI ethics issues among team members?

☐ Are there designated individuals responsible for AI ethics oversight?

Interdisciplinary Skills:

☐ Can your team effectively collaborate across disciplines (e.g., AI, domain expertise, project management)?

☐ Are there processes in place to facilitate interdisciplinary collaboration?

Project Management Readiness

AI-Specific Project Management:

☐ Do project managers have experience with AI-driven projects?

☐ Are there processes in place for managing the unique aspects of AI projects (e.g., data management, model iteration)?

Agile Methodologies:

☐ Is your team familiar with agile methodologies that are suitable for AI projects?

☐ Are there systems in place for rapid prototyping and iteration?

Risk Management:

☐ Have potential risks associated with AI projects been identified?

☐ Is there a risk management strategy specific to AI initiatives?

Performance Metrics:

☐ Have clear metrics been established for evaluating AI project success?

☐ Is there a system for ongoing monitoring and evaluation of AI performance?

Cultural Readiness for AI Adoption

Leadership Support:

☐ Is there strong leadership support for AI initiatives?

☐ Is there a clear vision for how AI fits into broader organizational goals?

Change Management:

☐ Is there a change management strategy for AI adoption?

☐ Are employees prepared for the changes AI might bring to their roles?

Innovation Culture:

☐ Is there a culture that supports experimentation and learning from failure?

☐ Are there processes for sharing AI lessons learned across the organization?

Ethical Considerations:

☐ Is there a framework for addressing ethical considerations in AI projects?

☐ Is there a process for ongoing ethical review of AI initiatives?

Stakeholder Buy-in:

☐ Have all relevant stakeholders been identified and engaged?

☐ Is there a communication plan for informing stakeholders about AI initiatives?

Legal and Compliance Readiness

AI Regulation Awareness:

☐ Is your team aware of relevant AI regulations in your industry and regions of operation?

☐ Is there a process for staying updated on evolving AI regulations?

Intellectual Property:

☐ Are there clear policies on IP ownership for AI-generated content?

☐ Have potential IP issues been considered and addressed?

Liability Considerations:

☐ Have potential liability issues for AI use been identified and addressed?

☐ Is appropriate insurance coverage in place for AI-related risks?

Contracting for AI:

☐ Are your contracting processes equipped to handle AI-specific considerations?

☐ Have you reviewed and updated relevant agreements (e.g., data sharing, model usage)?

Financial Readiness

Budget Allocation:

☐ Is there a dedicated budget for AI initiatives?

☐ Are there processes for evaluating the ROI of AI projects?

Cost Management:

☐ Is there an understanding of the potential ongoing costs associated with AI projects?

☐ Are there systems in place to monitor and manage AI-related expenses?

Investment Strategy:

☐ Is there a clear strategy for investing in AI capabilities over time?

☐ Are there contingency funds available to address unexpected AI-related challenges?

Working through this checklist, you can comprehensively understand your readiness for AI-driven projects. Remember, having every item checked off before starting is unnecessary – this assessment can also serve as a roadmap for areas to improve as you begin your AI journey. The following section will explore some case studies of successful AI implementations to provide practical insights into real-world applications.

10.5 Key Principles for Ethical and Effective AI Collaboration

As we've explored throughout this chapter, collaborating with AI models like Claude or ChatGPT can significantly enhance project outcomes. However, adhering to certain fundamental principles ensures that this collaboration is ethical and practical. This section will outline these principles and guide you through implementing them in your AI-driven projects.

Transparency and Accountability

Principle: Be open about using AI in your projects and establish clear lines of accountability for AI-driven decisions.

Implementation:

- Disclose when AI is being used in your processes or outputs.
- Document the role of AI in decision-making processes.
- Establish protocols for tracking and attributing AI contributions.
- Designate human oversight for critical AI-driven decisions.

Example: When using AI-generated content in marketing materials, include a note explaining that the content was created in collaboration with AI and specify the human review process it underwent.

Human-Centered AI Design

Principle: Design AI systems and workflows that augment and empower human capabilities rather than aiming to replace them.

Implementation:

- Involve end-users in the design process of AI systems.
- Focus on creating AI tools that enhance human decision-making and creativity.
- Prioritize user experience and intuitive interaction in AI interfaces.
- Provide options for human override in AI-driven processes.

Example: In a content creation workflow, use AI to generate ideas and initial drafts but keep human writing central to the process for refinement, fact-checking, and adding unique insights.

Bias Awareness and Mitigation

Principle: Actively work to identify, mitigate, and monitor for biases in AI systems and outputs.

Implementation:

- Conduct regular bias audits of your AI systems and their outputs.
- Use diverse datasets to train and test AI models.
- Implement bias detection tools in your AI workflows.
- Foster a culture of bias awareness among team members working with AI.

Example: In a recruitment AI system, regularly analyze the system's recommendations for potential gender or racial biases and adjust algorithms or datasets accordingly.

Data Privacy and Security

Principle: Protect the privacy of individuals and the security of data used in AI processes.

Implementation:

- Adhere to AI projects' data protection regulations (e.g., GDPR, CCPA).
- Implement strong data encryption and access controls. - Minimize data collection to what's necessary for the task at hand.
- Be transparent about data usage and provide opt-out options where appropriate.

Example: When using customer data to load into a product recommendation AI, ensure all data is anonymized, securely stored, and used only for the stated purpose.

Continuous Improvement and Adaptation

Principle: View AI implementation as an ongoing learning, improvement, and adaptation process.

Implementation:

- Establish feedback loops for AI system performance.
- Regularly shop for new AI models with more recent data, performance, and insights.
- Keep up-to-date with developments in AI technology and ethical considerations.
- Encourage an environment where team members are constantly learning and growing.

Example: In a customer service AI, continuously analyze customer feedback and agent corrections to improve the AI's responses over time.

Ethical Consideration at Every Stage

Principle: Integrate ethical considerations into every development and implementation stage of the AI project.

Implementation:

- Conduct ethical impact assessments at the outset of AI projects.
- Include ethics checkpoints in your project development lifecycle.
- Establish an AI ethics committee or designate ethics officers for AI projects.
- Provide ethics training for all team members involved in AI projects.

Example: Conduct a thorough ethical review before deploying an AI-driven credit scoring system to ensure it doesn't disadvantage particular demographic groups unfairly.

Interdisciplinary Collaboration

Principle: Encourage collaboration between AI specialists, domain experts, ethicists, and other relevant stakeholders.

Implementation:

- Form diverse teams for AI projects, including technical and non-technical members.
- Facilitate regular cross-functional meetings and knowledge-sharing sessions.
- Encourage a common language for discussing AI concepts across disciplines.
- Value and incorporate diverse perspectives in AI decision-making processes.

Example: In developing an AI system for medical diagnosis, ensure close collaboration between AI engineers, medical professionals, ethicists, and patient advocates throughout the process.

Explainability and Interpretability

Principle: Strive for AI systems and decisions humans can explain and interpret.

Implementation:

- Use explainable AI techniques where possible, especially for high-stakes decisions.
- Develop clear protocols for interpreting and communicating AI outputs.
- Provide training to relevant stakeholders on understanding AI decision processes.
- Be prepared to provide explanations for AI-driven decisions when required.

Example: In a financial AI system recommending investment strategies, ensure the system can provide clear explanations, including the key factors influencing each decision.

10.6 Conclusion: Embracing Continuous Evolution

As we wrap up our discussion on staying informed and evolving in the realm of AI, it's clear that achieving success demands a dedication to ongoing learning and flexibility. The strategies and frameworks we've covered establish a solid basis for continuous growth.

KEY INSIGHTS FROM THIS CHAPTER INCLUDE:

- The importance of maintaining current knowledge in the rapidly evolving AI landscape
- Strategies for effective integration of AI into existing workflows
- Tools for assessing and improving organizational AI readiness
- Principles for sustainable and ethical AI growth

Staying current with AI is not just about keeping up with the latest technologies—it's about thoughtfully evaluating how these advances can benefit your organization while maintaining ethical standards and operational effectiveness. The goal is not to implement every new AI capability but to selectively adopt those that align with your objectives and values.

As you continue your journey with AI, use the frameworks and strategies we've discussed as guideposts, but remain flexible and open to new approaches. AI will continue to evolve, and your success will depend on your ability to grow with it while focusing on your core objectives and ethical principles.

Looking ahead to the next chapter, we'll explore specific techniques for implementing ethical AI practices, building upon the growth mindset and organizational capabilities we've developed here.

CHAPTER 10 SUMMARY: STAYING CURRENT AND GROWING WITH AI

Key Concepts:

1. Strategies for staying informed about AI developments
2. Methods for integrating AI into existing workflows
3. Comprehensive AI readiness assessment
4. Core principles for sustainable AI growth
5. Future-focused implementation strategies

Reflection Questions:

1. How do you plan to keep pace with rapid AI developments in your field?
2. What obstacles do you foresee in integrating AI into your current workflows?
3. How will you measure your organization's AI readiness and progress?

Action Items:

1. Create a personal AI learning roadmap
2. Develop an AI integration plan for your existing workflows
3. Conduct an AI readiness assessment for your organization

Further Exploration:

- Research emerging AI trends in your industry
- Study successful AI integration case studies
- Investigate AI readiness assessment frameworks

11

IMPLEMENTING ETHICAL AI: TECHNIQUES AND STRATEGIES FOR RESPONSIBLE DEVELOPMENT

11.1 Introduction: The Ethical Imperative in AI Collaboration

In this book, we've discussed how partnering with AI can transform our work and creative processes. Yet, with this advancement comes the need for careful consideration. This chapter will provide actionable strategies for incorporating ethical AI practices into your projects, building on our insights about AI language models in Chapter 2 and conversation techniques in Chapter 4.

In this chapter, we'll explore the core elements of ethical AI, share practical strategies to tackle prevalent challenges, and present exercises to enhance your ethical AI proficiency. By the end, you'll be equipped with tools to ensure your AI projects are impactful, responsible, and contribute positively to society.

11.2 Understanding the Ethical AI Landscape

Before we discuss specific techniques, let's examine the key components of ethical AI, much like we did when dissecting AI language models in Chapter 2.

The Four Pillars of Ethical AI

Fairness: Ensuring AI systems treat all individuals and groups equitably.

Transparency: Making AI decision-making processes understandable and explainable.

Privacy: Protecting individual data rights and preventing unauthorized use.

Accountability: Establishing clear responsibility for AI actions and decisions.

These pillars form the foundation of ethical AI practice. Let's explore each in detail:

Fairness

Fairness in AI involves ensuring that AI systems operate without bias or discrimination. This is important because biased algorithms can reinforce and even exacerbate societal disparities.

Key Concepts:

Disparate Impact: Disparate Impact occurs when an AI system's outcomes disproportionately affect certain groups, even if they were not intentionally designed to do so.

Equality vs. Equity: Equality means treating everyone equally, while equity involves accounting for historical disadvantages to ensure fair outcomes.

Transparency

Transparency involves making AI systems understandable to both developers and end-users. It's essential for building trust and enabling meaningful oversight.

Key Concepts:

Explainable AI (XAI): Methods that enable individuals to comprehend and analyze the decisions made by AI systems.

Model Cards: Documentation that provides valuable information about a model's performance, limitations, and intended use.

Privacy

Privacy in AI focuses on protecting individual data rights and preventing misuse of personal information.

Key Concepts:

Data Minimization: Collecting and using only the data necessary for the task.

Differential Privacy: A privacy method that adds calculated noise to data, making individuals unidentifiable while keeping statistics functional.

Accountability

Accountability involves establishing clear responsibility for AI systems' actions and decisions.

Key Concepts:

Human-in-the-Loop: Keeping humans involved in critical decision-making processes.

Audit Trails: Maintaining detailed records of an AI system's decision-making process.

11.3 Practical Techniques for Ethical AI Implementation

Having grasped the essential elements, let's explore practical methods for implementing ethical AI, similar to our discussion of AI conversation strategies in Chapter 4.

Fairness Techniques

Diverse Data Collection: Use stratified sampling to ensure your training data represents diverse populations.

> "Please analyze our current dataset for the hiring AI system. Identify any underrepresented groups and suggest strategies for collecting additional data to ensure fair representation across all demographics."

Bias Detection and Mitigation: Regularly audit your AI system for bias and implement corrective measures.

> "Conduct a bias audit on our loan approval AI system. Specifically, check for any disparities in approval rates across different ethnic groups and genders. If biases are found, suggest mitigation strategies."

Fairness-Aware Machine Learning: Incorporate fairness constraints into your AI model's training process.

> "Explain how we can implement fairness constraints in our hiring AI model. What types of constraints would be most effective in ensuring equal opportunity across all applicant groups?"

Transparency Techniques

Explainable AI Methods: Implement methods that provide clear explanations for AI decisions.

> "For our medical diagnosis AI, we must make its decision-making process more transparent. Suggest three explainable AI techniques we could implement, and explain how each would help doctors understand the AI's reasoning."

Interpretable Model Design: Design AI models with interpretability in mind from the start

> "We're developing a stock market"AI" assistant. How can we design the model to be inherently interpretable? What types of model architectures or features would allow for easier explanation of its decisions to customers?"

Model Cards: Create comprehensive documentation for your AI models.

> "Generate a model card for our customer churn prediction AI. Include details on its intended use, performance metrics, ethical considerations, and limitations. What approach can we take to convey this information clearly to both technical and non-technical audiences?"

Privacy Techniques

Data Minimization: Collect and use only the data necessary for the task.

> "Review the data collection process for our personalized shopping recommendation AI. Identify any data points we're collecting that aren't essential for generating accurate recommendations. How can we minimize data collection while maintaining system performance?"

Differential Privacy: Implement differential privacy to protect individual data while maintaining overall accuracy.

> "Explain how we can apply differential privacy to our health data analysis AI. What are the trade-offs between privacy protection and accuracy? How should we determine the appropriate privacy budget?"

Federated Learning: Train AI models on decentralized data without sharing raw data.

> "Our mobile keyboard prediction AI must learn from user data without compromising privacy. Describe how we can implement federated learning in this context. What challenges might we face, and how can we address them?"

Accountability Techniques

Audit Trails: Implement comprehensive logging for AI decision-making.

> "Design an audit trail system for our autonomous vehicle AI. What key information should we log for each decision made by the AI? How can we ensure this logging system is tamper-proof and easily accessible for investigations?"

Human-in-the-Loop Systems: Implement approval workflows for critical AI decisions.

> "We need to incorporate very strict human oversight for our AI-assisted judicial recommendation system. Describe a human-in-the-loop process that allows for AI recommendations while ensuring final decisions are made by human judges with adequate context and defensive mitigations for biases. How can we design this system to prevent over-reliance on AI suggestions?"

11.4 Practical Exercises: Developing Your Ethical AI Skills

Let's put these techniques into practice with some hands-on exercises similar to those in Chapter 5.

Exercise 1: Fairness Audit

Goal: Conduct a fairness audit on a hypothetical AI hiring system.

Steps:

1. Imagine you're working with an AI hiring system that's been used for six months. You've noticed that the system favors certain demographic groups in its recommendations.
2. Write a prompt asking the AI to analyze its decision-making patterns for potential biases. How would you phrase this request to get the most comprehensive and honest assessment?
3. Based on the AI's response, draft a plan for aggressively mitigating any biases found. Include specific changes to the data collection process, model architecture, and decision thresholds.
4. Create a prompt that asks the AI how to implement each of these changes.
5. Finally, a monitoring system should be designed to continuously check for fairness in the AI's future hiring recommendations.

Exercise 2: Explainable AI Dashboard

Goal: Create a conceptual design for an explainable AI dashboard for a medical diagnosis system.

Steps:

1. Envision an AI system that assists doctors in diagnosing complex diseases. Your task is to create a dashboard that explains the AI's diagnostic reasoning to doctors and patients.
2. Write a prompt asking the AI to suggest key components this dashboard should include. Consider what information would be most crucial for doctors versus patients.
3. Based on the AI's suggestions, sketch out a rough layout for the dashboard. Include areas for displaying confidence levels, key factors influencing the diagnosis, and potential alternative diagnoses.
4. Create a prompt asking the AI how it would generate plain-language explanations of its diagnostic process for patients. How would these explanations differ from the more technical ones provided to doctors?
5. Design a feature that allows doctors to ask follow-up questions about the AI's reasoning. Write example prompts that a doctor might use to probe deeper into the AI's diagnostic process.

Exercise 3: Privacy-Preserving Machine Learning

Goal: Develop a strategy for implementing privacy-preserving techniques in a personal finance AI advisor.

Steps:

1. Imagine creating an AI that tailors financial advice to individuals by analyzing their spending patterns, income levels, and specific economic objectives.
2. Write a prompt to the AI asking it to explain the potential privacy risks associated with collecting and analyzing personal financial data.
3. Create a data minimization strategy Based on the AI's response. What is the minimum amount of data you need to provide adequate financial advice?
4. Develop a prompt asking the AI how to implement differential privacy in this context. How would you balance privacy protection with the accuracy of financial recommendations?
5. Design a user interface that communicates to users what data is being collected, how it's being used, and what privacy protections are in place. Write prompts to the AI for help in crafting clear, user-friendly explanations of complex privacy concepts.

11.5 Case Study: Ethical AI in Action

Let's apply these techniques to a real-world scenario: developing an AI-driven financial product recommendation system for a financial institution.

Step 1: Ethical Impact Assessment

Begin by performing a comprehensive evaluation of possible ethical concerns. Create a prompt for the AI to help identify potential ethical concerns:

> "We're developing an AI-driven financial product recommendation system. Please help us identify potential ethical issues in fairness, transparency, privacy, and accountability. For each issue, suggest preliminary mitigation strategies."

Based on the AI's response, create a comprehensive list of ethical considerations and initial strategies to address them.

Step 2: Data Collection and Preparation

Implement strategies to ensure diverse and representative data. Prompt the AI for guidance:

> "How can we ensure our training data for the financial product recommendation AI is diverse and representative of all demographic groups? What sampling techniques or data augmentation methods would you recommend?"

Use the AI's suggestions to develop a robust data collection and preparation strategy.

Step 3: Model Development with Fairness Constraints

Develop the financial product recommendation model with built-in fairness constraints. Consult the AI on implementation:

> "We want to incorporate fairness constraints into our financial product recommendation model. Can you explain the different types of fairness metrics we should consider? How can we balance these fairness considerations with the model's predictive accuracy?"

Based on the AI's advice, design a model development process that prioritizes fairness and accuracy.

Step 4: Implementing Explainability

Ensure the financial product recommendation system can explain its decisions. Prompt the AI for ideas:

> "We need our financial product recommendation AI to explain its decisions clearly to our loan officers and applicants. What explainable AI techniques would you recommend? How should we adapt these explanations for technical and non-technical audiences?"

Use the AI's suggestions to implement a multi-layered explanation system for your financial product recommendation AI.

Step 5: Privacy Protection

Implement strong privacy protections for applicant data. Consult the AI on best practices:

> "How can we implement robust privacy protections in our financial product recommendation system? Please suggest techniques for data anonymization, encryption, and access control. How can we ensure compliance with relevant data protection regulations?"

Develop a comprehensive privacy protection strategy based on the AI's recommendations.

Step 6: Accountability Measures

Establish clear accountability measures for the AI system. Prompt the AI for guidance:

> "We need to ensure clear accountability for our financial product recommendation AI. How should we implement audit trails? What kind of human oversight is necessary, and at what stages of the decision-making process?"

Design an accountability framework that includes detailed logging, regular audits, and appropriate human oversight.

Step 7: Ethical Review and Iteration

Finally, establish a process for ongoing ethical review and improvement. Ask the AI:

> "How can we set up a regular ethical review process for our financial product recommendation AI? What metrics should we track to ensure ongoing fairness, transparency, privacy protection, and accountability?"

Use the AI's suggestions to create a robust ethical review process, ensuring your financial product recommendation system remains ethically sound as it evolves.

11.6 Conclusion: Ethical AI as a Competitive Advantage

By implementing these ethical AI techniques, you're building more robust, trustworthy AI systems that can provide long-term value. Remember, ethical AI is an ongoing process that requires constant vigilance and adaptation.

Keep these ethical considerations at the forefront of your work as you continue your AI collaboration journey. By doing so, you'll create more responsible AI systems and gain a competitive edge in an increasingly AI-driven world.

Ethical AI should be viewed as a driving force for innovation, fostering the creation of AI systems that prioritize reliability, trustworthiness, and alignment with human values. As the technology evolves, the importance of developing and implementing ethical AI will grow, making it a critical skill for the future.

In our final chapter, we'll look ahead to the future of human-AI collaboration and explore how you can prepare for the exciting developments on the horizon, building on the ethical foundation above.

CHAPTER 11 SUMMARY: IMPLEMENTING ETHICAL AI: TECHNIQUES AND STRATEGIES FOR RESPONSIBLE DEVELOPMENT

Key Concepts:

1. The Four Pillars of Ethical AI: Fairness, Transparency, Privacy, and Accountability
2. Practical techniques for implementing ethical AI in real-world scenarios
3. Prompt-based approaches to addressing ethical concerns in AI systems
4. Strategies for conducting ethical impact assessments and ongoing ethical reviews

Reflection Questions:

1. How might you apply the fairness techniques discussed in this chapter to an AI project in your field?
2. What obstacles do you anticipate when integrating explainable AI into intricate decision-making systems?
3. How does the concept of privacy-preserving machine learning impact your approach to AI development?

Action Items:

1. Conduct an ethical impact assessment for a current or planned AI project in your organization
2. Develop a set of prompts for auditing an AI system's fairness and transparency
3. Create a model card for an AI system you're familiar with, focusing on ethical considerations

Further Exploration:

- Research emerging standards and best practices for ethical AI in your industry
- Explore case studies of successful ethical AI implementations in various fields
- Investigate tools and frameworks for ongoing ethical assessment of AI systems

12

ADVANCED TECHNIQUES FOR FUTURE HUMAN-AI COLLABORATION

12.1 Introduction: Preparing for Next-Generation AI Systems

As we look to the future of human-AI collaboration, it's crucial to develop advanced techniques that will allow us to work effectively with increasingly sophisticated AI systems. In this chapter, we'll explore practical strategies for mastering emerging AI technologies, like we examined AI language models in Chapter 2 and conversation strategies in Chapter 4.

We'll break down key emerging technologies, provide hands-on techniques for leveraging them, and offer practical exercises to help you develop the skills needed for future AI collaboration. By the conclusion of this chapter, you'll be equipped with essential tools to navigate the fast-changing world of AI effectively.

12.2 Mastering Multimodal AI Interaction

Multimodal AI systems, which can process and generate content across various formats (text, image, audio, video), represent the next frontier in human-AI collaboration. Let's explore techniques for effectively working with these advanced systems.

Understanding Multimodal AI Architecture

Multimodal AI systems typically consist of:

Input Processing Modules: Separate neural networks for processing each type of input (e.g., Convolutional Neural Networks for images, Transformers for text).

Fusion Layer: A mechanism for combining information from different modalities.

Joint Representation Space: Where the fused information is processed.

Output Generation Modules: These produce various types of output.

Techniques for Effective Multimodal Prompting

Complementary Information: Provide information in multiple modalities that complement each other rather than repeating the same information.

> [Image of a complex mechanical system]
>
> "Explain how this system works, focusing on the parts labeled A, B, and C. Generate a text description and create a simplified diagram highlighting the key components and their interactions."

Cross-Modal Grounding Use one modality to provide context or constraints for another.

> "Generate a realistic image of a futuristic city based on the following description: [detailed text description]. Then, provide a text analysis of how the generated image reflects or diverges from key elements in the original description."

Sequential Multimodal Interaction Build up complex outputs through a series of multimodal interactions.

1. "Create a basic 3D model of a smartphone based on this text description: [description]"
2. [User uploads a sketch] "Modify the 3D model to incorporate elements from this hand-drawn sketch."
3. "Generate a voiceover script explaining the key features of this smartphone design."
4. "Based on the 3D model and voiceover script, create a 30-second video advertisement for this smartphone."

Exercise: Multimodal AI Collaboration

Goal: Design a product using multimodal AI assistance.

Steps:

1. Write a text description of a novel product idea.
2. Use a multimodal AI to generate a rough sketch based on your description.
3. Provide verbal feedback to refine the sketch, using specific references to parts of the image.
4. Ask the AI to generate a list of potential features based on the refined sketch.
5. Create a prompt that combines the sketch, feature list, and a request for a marketing tagline.

12.3 Leveraging Advanced Natural Language Processing

As NLP models become more sophisticated, new techniques are needed to harness their capabilities thoroughly. Let's explore some advanced strategies:

Multi-Turn Reasoning

Guide the AI through a step-by-step reasoning process for complex problems.

> "We need to solve a complex logical puzzle. Let's approach this step-by-step:
>
> 1. First, list all the given facts from the puzzle.
>
> 2. Identify any potential inconsistencies or ambiguities in these facts.
>
> 3. Propose three possible logical approaches to solving the puzzle.
>
> 4. For each approach, walk through the reasoning steps, highlighting any assumptions made.
>
> 5. Compare the results from each approach and determine the most likely correct solution.
>
> 6. Explain why this solution is the most probable, addressing any potential weaknesses in the reasoning."

Hypothetical Scenario Exploration

Use the AI to explore multiple hypothetical scenarios to gain deeper insights.

> "We're developing a new urban transportation system. Let's explore three scenarios:
>
> 1. Scenario A: Fully autonomous electric vehicles
>
> 2. Scenario B: Underground hyperloop system
>
> 3. Scenario C: Elevated personal rapid transit pods
>
> For each scenario:
>
> a) Describe the critical technologies required
>
> b) Analyze potential societal impacts (economic, environmental, social)
>
> c) Identify significant implementation challenges
>
> d) Propose solutions to these challenges
>
> After exploring all scenarios, synthesize the insights to propose a hybrid solution combining each approach's best elements."

Adversarial Prompting

Intentionally challenge the AI's outputs to refine and improve them.

1. "Provide a comprehensive argument for implementing a universal basic income."
2. "Now, act as a critic and identify the three weakest points in the argument you just presented."
3. "As the original proponent, rebut these criticisms, providing additional evidence or reasoning."
4. "Now, step back and analyze this debate. What are the underlying assumptions on both sides? Are there any logical fallacies present?"
5. "Based on this analysis, provide a refined, more nuanced argument about universal basic income that addresses the complexities revealed through this process."

Exercise: Advanced NLP Problem-Solving

Goal: Use advanced NLP techniques to develop a complex strategy.

Steps:

1. Choose a complex problem in your field (e.g., reducing urban pollution, improving online education).
2. Use the multi-turn reasoning technique to break down the problem and analyze its components.
3. Apply the hypothetical scenario exploration technique to consider at least three different approaches to solving the problem.
4. Use adversarial prompting to challenge and refine the best solution from step 3.
5. Synthesize the insights gained from this process into a comprehensive strategy document.

12.4 Implementing Next-Generation Explainable AI (XAI)

As AI systems become more complex, advanced explainability techniques are crucial for maintaining transparency and trust. Let's explore cutting-edge XAI methods:

Layered Explanations

Generate explanations at multiple levels of detail and technicality.

> "Explain the decision-making process of an AI trading algorithm that just made a significant stock purchase. Provide explanations at three levels:
>
> 1. Executive Summary: A brief, non-technical explanation suitable for senior management.
>
> 2. Analyst Level: A more detailed explanation, including key financial metrics and market trends considered.
>
> 3. Technical Deep Dive: A comprehensive explanation including the specific AI models used, key features in the data, and how they influenced the decision."

Counterfactual Explanations

Explain AI decisions by showing how different inputs would change the outcome.

> "Our AI system denied a loan application. Using counterfactual explanations:
>
> 1. Identify the top three factors that led to the denial.
>
> 2. For each factor, explain what change would be necessary to flip the decision to an approval.
>
> 3. Generate a visual representation (like a decision tree or influence diagram) showing how these factors interact and influence the decision.
>
> 4. Provide a natural language explanation of this counterfactual analysis that could be given to the loan applicant."

Feature Visualization

Create visual representations of the features and patterns the AI detects.

> "For an AI system that detects potential fraud in financial transactions:
>
> 1. Generate a heatmap showing the relative importance of different transaction features in the decision-making process.
>
> 2. Create a series of example transactions that gradually transition from 'legitimate' to 'highly suspicious,' explaining how the AI's assessment changes with each step.
>
> 3. Produce a network diagram showing how different features interact and influence each other in the AI's decision-making process."

Exercise: Designing an XAI Interface

Goal: Create a user interface to explain complex AI decisions to non-technical users.

Steps:

1. Choose a complex AI system (e.g., a medical diagnosis AI or a financial robo-advisor).
2. Use the layered explanations technique to create three levels of explanation for the system's decisions.
3. Implement counterfactual explanations to show users how different inputs would affect the outcome.
4. Design a feature visualization that intuitively displays the critical factors in the AI's decision-making.
5. Prototype an interface that combines these elements, allowing users to interactively explore the AI's decision-making process.

12.5 Advanced AI Ethics: Future-Facing Considerations

Based on the ethical framework discussed in Chapter 11, advanced AI systems present distinct challenges that necessitate targeted strategies. This section focuses on the new ethical considerations arising from next-generation AI technologies' capabilities.

Advanced Impact Assessment Techniques

Advanced AI systems require enhanced assessment methods beyond traditional ethical frameworks. Here's how to adapt impact assessments for sophisticated AI:

> "For a next-generation AI hiring system with autonomous learning capabilities:
>
> 1. Model potential second-order effects (e.g., how might the system's decisions influence industry-wide hiring practices?)
>
> 2. Assess emergent behavioral risks unique to learning systems
>
> 3. Evaluate potential societal impacts beyond immediate stakeholders
>
> 4. Design safeguards specifically for autonomous decision evolution
>
> 5. Create monitoring systems capable of detecting unexpected behavioral patterns."

Future-Proofing Value Alignment

As AI systems become more sophisticated, traditional value alignment methods need enhancement. This section explores techniques for maintaining ethical alignment as systems evolve:

1. "How might this AI system's understanding of its ethical principles evolve as it learns? Design guardrails for ethical concept drift."
2. "Create mechanisms to detect when the system's operational context has shifted significantly enough to require ethical framework updates."
3. "Develop protocols for safely testing ethical boundaries in novel situations while maintaining core principles."
4. "Design feedback systems that allow the AI to strengthen its ethical reasoning without compromising fundamental values."

Next-Generation Transparency

Advanced AI systems require new approaches to transparency that account for increasingly complex decision-making:

> "For an AI system with dynamic learning capabilities:
>
> 1. Design methods to track and explain evolutionary changes in the system's decision-making
>
> 2. Create visualization tools for complex ethical trade-offs
>
> 3. Develop real-time ethical decision-monitoring dashboards
>
> 4. Implement stakeholder-specific transparency layers that evolve with the system
>
> 5. Design protocols for explaining unexpected ethical decisions."

Exercise: Future Ethics Planning

Goal: Prepare your organization for emerging ethical challenges in advanced AI systems.

Steps:

1. Identify potential ethical challenges unique to learning/evolving AI systems
2. Design monitoring systems capable of detecting emerging ethical concerns
3. Create protocols for managing unexpected ethical scenarios
4. Develop frameworks for ethical oversight of autonomous learning
5. Plan for regular updates to ethical guidelines as AI capabilities advance

12.6 Preparing for Artificial General Intelligence (AGI)

While AGI remains a prospect, developing techniques for potential collaboration with highly advanced AI systems is crucial. These techniques can also enhance our work with current AI:

Meta-Learning Prompts

Design prompts encouraging AI to improve its learning and problem-solving strategies.

> "We're going to work on improving your problem-solving capabilities:
>
> 1. Describe your current approach to solving complex, multi-step problems.
>
> 2. Analyze the strengths and weaknesses of this approach.
>
> 3. Research three alternative problem-solving methodologies humans use (e.g., design thinking, systems thinking, first principles reasoning).
>
> 4. For each methodology, explain how you could incorporate its critical principles into your problem-solving process.
>
> 5. Propose a new, hybrid problem-solving approach that combines the best elements of your current method and these human methodologies.
>
> 6. Design a series of increasingly complex test problems to evaluate this new approach.
>
> 7. After working through these test problems, reflect on the effectiveness of the new approach and suggest further refinements."

Cognitive Architecture Alignment

Align AI cognitive processes more closely with human cognitive architectures to improve collaboration.

"Let's explore aligning your cognitive processes more closely with human cognition:

1. Summarize critical aspects of human cognitive architecture (e.g., working memory, long-term memory, attention mechanisms).

2. Describe how your cognitive processes are similar or different for each aspect.

3. Propose modifications to your cognitive processes that could make them more analogous to human cognition. Consider trade-offs between capability and alignment.

4. Design a series of collaborative tasks that would test the effectiveness of these modified cognitive processes in human-AI teamwork.

5. Based on hypothetical results from these tasks, suggest further refinements to the cognitive alignment process."

Ethical Boundary Setting

Establish clear ethical boundaries and decision-making protocols for advanced AI systems.

> "We need to establish ethical boundaries for an advanced AI system:
>
> 1. Define a core set of ethical principles (e.g., beneficence, non-maleficence, autonomy, justice).
>
> 2. For each principle, provide specific behavioral guidelines and decision-making rules.
>
> 3. Create a hierarchical ethical decision-making framework for resolving conflicts between principles.
>
> 4. Design a series of complex ethical dilemmas to test this framework.
>
> 5. Based on the outcomes of these dilemmas, refine the ethical boundaries and decision-making protocols.
>
> 6. Develop a system for explaining ethical decisions made by the AI, including the principles invoked and the reasoning process.
>
> 7. Propose a mechanism for ongoing human oversight and adjustment of these ethical boundaries as the AI system evolves."

Exercise: AGI Collaboration Scenario Planning

Goal: Develop strategies for effective collaboration with a hypothetical AGI system.

Steps:

1. Define the capabilities and limitations of a hypothetical AGI system.
2. Use meta-learning prompts to design a learning curriculum for this AGI to enhance its collaboration skills.
3. Apply cognitive architecture alignment techniques to identify potential challenges and opportunities in human-AGI teamwork.
4. Establish ethical boundaries and decision-making protocols for the AGI system.
5. Create a detailed scenario of a complex project undertaken collaboratively by a human team and the AGI system, highlighting key interaction points and potential issues.
6. Develop strategies for addressing the challenges identified in this scenario.

12.7 Conclusion: Evolving with AI

As we've explored in this chapter, the future of human-AI collaboration will require us to evolve our skills and techniques continuously. To be leaders in this fast-evolving field, we must focus on several key areas: mastering multimodal interactions, utilizing advanced natural language processing, adopting next-generation explainable AI, unwaveringly prioritizing ethical considerations, and gearing up for more sophisticated AI systems.

Remember that these advanced techniques build upon the foundational skills we've developed throughout this book. As we learned to guide AI attention in earlier chapters, we now better understand how to navigate complex multimodal interactions. Our understanding of prompt engineering evolves into sophisticated multi-turn reasoning and meta-learning strategies.

As AI advances, our role shifts towards becoming skilled conductors of increasingly capable AI ensembles. We must balance leveraging AI capabilities with maintaining human oversight, creativity, and ethical judgment.

The partnership between humans and AI will shape the future, emphasizing collaboration over conflict. As a practitioner of AI in your profession, you're playing a crucial role in shaping that future and addressing intricate challenges together.

Continue to practice these techniques, stay informed about AI advancements, and always approach AI collaboration with curiosity, critical thinking, and ethical consideration. The journey of discovery in human-AI partnership is just beginning, and I'm excited to see where you'll take it next.

CHAPTER 12 SUMMARY: ADVANCED TECHNIQUES FOR FUTURE HUMAN-AI COLLABORATION

Key Concepts:

1. Multimodal AI interaction strategies
2. Advanced Natural Language Processing techniques
3. Next-generation Explainable AI (XAI) methods
4. Ethical considerations in advanced AI systems
5. Preparation for potential Artificial General Intelligence (AGI)

Reflection Questions:

1. How might you apply multimodal AI techniques to enhance collaboration in your field?
2. What obstacles do you anticipate in adopting advanced NLP techniques?
3. How does the concept of next-generation XAI impact your approach to AI transparency and trust-building?

Action Items:

1. Design a multimodal AI interaction for a complex task in your domain
2. Practice using multi-turn reasoning and adversarial prompting techniques
3. Develop an ethical impact assessment for an AI system in your organization

Further Exploration:

- Research cutting-edge multimodal AI systems and their applications
- Explore recent advancements in AI ethics and value alignment techniques
- Investigate current approaches to AGI development and potential implications for human-AI collaboration

APPENDIX A: GLOSSARY OF TERMS

Artificial Intelligence (AI): The processes by which machines, particularly computer systems, imitate human intelligence.

Machine Learning (ML): Field of research where computer systems can get better at specific tasks by analyzing patterns in data. The more examples they process, the more accurate they become - like a student learning from practice.

Deep Learning: Deep learning uses layered networks that process information similarly to brain cells. Each layer extracts increasingly complex patterns from the data, building up from fundamental features to sophisticated concepts.

Natural Language Processing (NLP): Natural language processing helps computers understand and respond to everyday human speech and writing. It bridges the gap between how machines process information and how people naturally communicate.

Large Language Model (LLM): Large language models are AI systems that learn patterns from billions of text examples, allowing them to read, write, and understand language like humans do. They absorb the structure and meaning of language from books, articles, and websites.

Prompt Engineering: Designing and refining inputs to AI systems to elicit desired outputs.

Context Window: The amount of preceding text an AI model can consider when generating a response.

Attention Mechanism: Attention mechanisms help AI systems prioritize what's relevant, similar to how humans focus on key details in a scene or conversation. The system learns which parts of the input matter most for each specific task.

Retrieval-Augmented Generation (RAG): A technique that combines information retrieval with text generation to produce more accurate and informed responses.

Fine-tuning: Fine-tuning customizes a general-purpose AI model for specific work - like teaching a math expert to focus on solving physics problems. The model builds on its broad knowledge while adapting to new, specialized tasks.

Transfer Learning: Transfer learning builds on existing AI knowledge, such as how a chef who masters French cooking can learn Italian cuisine faster. Rather than starting from scratch, the AI applies its foundational skills to master new, related tasks.

Transformer: A type of deep learning model architecture that has become the foundation for many state-of-the-art NLP systems.

Tokenization: Tokenization splits text into digestible pieces for AI - like breaking a book into words, parts of words, or even single letters. This lets the system process language in consistent, manageable chunks.

Hallucination: In AI, it refers to when a model generates information that is not accurate or grounded in its training data.

Explainable AI (XAI): AI systems that make their decision-making processes transparent and understandable to humans.

APPENDIX B: RESOURCES AND FURTHER READING

Books

"Artificial Intelligence: A Modern Approach" by Stuart Russell and Peter Norvig

"Deep Learning" by Ian Goodfellow, Yoshua Bengio, and Aaron Courville

"The Hundred-Page Machine Learning Book" by Andriy Burkov

"AI Ethics" by Mark Coeckelbergh

Online Courses

Coursera - "AI for Everyone" by Andrew Ng

edX - "Artificial Intelligence (AI)" by Columbia University

Fast.ai - "Practical Deep Learning for Coders"

Google AI - "Machine Learning Crash Course"

Websites and Blogs

OpenAI Blog (https://openai.com/blog/)

Google AI Blog (https://ai.googleblog.com/)

Towards Data Science (https://towardsdatascience.com/)

Arxiv Sanity Preserver (http://www.arxiv-sanity.com/)

Podcasts

"Lex Fridman Podcast" - interviews with AI researchers and practitioners

"AI in Business" - focuses on practical applications of AI in various industries

"The TWIML AI Podcast" - This Week in Machine Learning & AI

Research Papers

"Attention Is All You Need" by Vaswani et al. (2017)

"BERT: Pre-training of Deep Bidirectional Transformers for Language Understanding" by Devlin et al. (2018)

"Language Models are Few-Shot Learners" (GPT-3 paper) by Brown et al. (2020)

Tools and Platforms

TensorFlow (https://www.tensorflow.org/)

PyTorch (https://pytorch.org/)

Hugging Face (https://huggingface.co/)

OpenAI Playground (https://beta.openai.com/playground)

APPENDIX C: FREQUENTLY ASKED QUESTIONS

Q: What's the difference between AI, machine learning, and deep learning?

A: AI is the broadest term for machines mimicking human cognitive functions. Machine learning systems improve at tasks through practice and exposure to examples, similar to how humans gain expertise. The more data they analyze, the better they become at spotting patterns and making decisions. Within this field, deep learning represents a specialized area that employs multiple layers of neural networks to process data more effectively.

Q: How do I choose the AI model for my task?

A: Consider factors like the specific task requirements, the amount and type of data available, computational resources, and the level of explainability needed. Starting with a pre-trained model and fine-tuning it can be effective for many tasks.

Q: What are the ethical concerns surrounding AI?

A: Key ethical concerns include bias in AI systems, privacy issues, job displacement, the potential for AI to be used maliciously, and the long-term implications of competent AI systems.

Q: How can I ensure my use of AI is ethical?

A: Stay informed about AI ethics, be transparent about AI use, regularly audit AI systems for bias, prioritize privacy and data protection, and maintain human oversight in critical decision-making processes.

Q: What's the future of human-AI collaboration?

A: The future likely involves a more seamless integration of AI into various aspects of work and life, with AI augmenting human capabilities rather than replacing them. This could include more advanced natural language interfaces, AI-assisted creativity tools, and AI systems that explain their reasoning and decision-making processes.

Q: How can I stay updated with the rapidly evolving field of AI?

A: Follow reputable AI blogs and news sources, attend AI conferences and workshops, participate in online AI communities, and regularly experiment with new AI tools and techniques.

Q: What skills should I develop to work effectively with AI?

A: Key skills include programming (especially Python), data analysis, understanding of machine learning concepts, prompt engineering, and domain expertise in your specific field. Soft skills like critical thinking, creativity, and ethical reasoning are also crucial.

APPENDIX D: COMMUNITY AND SUPPORT

Online Communities

Reddit - r/artificial, r/MachineLearning, r/learnmachinelearning

Stack Overflow - tags: artificial intelligence, machine learning

Kaggle - forums and competitions for data science and machine learning

Conferences

NeurIPS (Neural Information Processing Systems)

ICML (International Conference on Machine Learning)

ICLR (International Conference on Learning Representations)

AI for Good Global Summit

Meetups and Local Groups

- Check Meetup.com for AI and machine learning groups in your area
- Look for local chapters of organizations like Women in Machine Learning and Data Science (WiMLDS)

Professional Organizations

Association for the Advancement of Artificial Intelligence (AAAI)

IEEE Computational Intelligence Society

Association for Computing Machinery (ACM) **Special Interest Group on Artificial Intelligence** (SIGAI)

Support Channels

- Many AI tools and platforms have their support forums or Discord channels
- GitHub repositories of open-source AI projects often have active issue trackers and discussion forums

> ***Remember:*** *The field of AI is rapidly evolving, so it's essential to regularly check these resources for the most up-to-date information and to actively participate in the community to stay current with the latest developments and best practices.*

APPENDIX E: AI CONTENT GENERATION AND EDITING TOOLS

This book was written with the assistance of:

Claude 3.5 Sonnet. Anthropic, 2024. https://anthropic.com

GPT-4o mini. OpenAI, 2024. https://openai.com

Grammarly. Grammarly, Inc., 2024. https://grammarly.com

o1-mini. OpenAI, 2024. https://openai.com

o1-preview. OpenAI, 2024. https://openai.com

APPENDIX F: REFERENCES

Asad, M., Shaukat, S., Javanmardi, E., Nakazato, J., & Tsukada, M. (2023). A comprehensive survey on privacy-preserving techniques in federated recommendation systems. *Applied Sciences*, 13(10), 6201.

Grimm, P., Grossman, M., & Cormack, G. (2021). Artificial Intelligence as evidence. *Northwestern Journal of Technology and Intellectual Property*, 19(1), 9-106.

Hassan, M., Aziz, L. A. R., & Andriansyah, Y. (2023). The role artificial intelligence in modern banking: An exploration of AI-driven approaches for enhanced fraud prevention, risk management, and regulatory compliance. Retrieved from https://core.ac.uk/download/578755756.pdf

Paltiel, M., Cheong, M., Coghlan, S., & Lederman, R. (2022). Teaching digital ethics in information systems. Retrieved from https://core.ac.uk/download/551321396.pdf

Pujol, D. S. (2022). Application of machine learning in models. https://doi.org/10.26524/royal.165

Romppanen, V. (2023). Deep neural networks for jamming and interference classification in 5G physical layer. Retrieved from https://core.ac.uk/download/595499385.pdf

Note: The following web resources were retrieved on October 22, 2024:

AI is a Tool. (2024). High risk AI: A deep dive into cybersecurity implications. https://aiisatool.com/ai-faqs/what-is-high-risk-ai-examples/

Alpha Hub. (2024). https://alphanumeric.dev/

Asana. (2024). "The Eisenhower Matrix: How to prioritize your to-do list." https://asana.com/resources/eisenhower-matrix

ASRYK. (2024). Scale AI: Building the scaffolding for tomorrow's artificial intelligence. https://asryk.com/scale-ai-building-the-scaffolding-for-tomorrows-artificial-intelligence/

Bostrom, N. (2003). "Ethical Issues in Advanced Artificial Intelligence." Philosophy Faculty, Oxford University. https://nickbostrom.com/ethics/ai.html

Deep Learning. (2024). Generative AI Commons. https://genaicommons.org/glossary/deep-learning/

DigitrendZ. (2024). AI. https://digitrendz.blog/glossary-archive/ai/

Herranz, M. (2023, July 14). The 4 pillars of Ethical AI – and why they're important to Machine Learning. Pangeanic Blog. https://blog.pangeanic.com/the-4-pillars-of-ethical-ai-and-why-theyre-important-to-machine-learning

Klu.ai. (2024). What is responsible AI? Klu AI Glossary. https://klu.ai/glossary/responsible-ai

Mind Global. (2024). Machine learning. https://www.mindglobal.ae/Machine-learning.php

OTSM-TRIZ. (2024). "OTSM-TRIZ Tools." https://sites.google.com/site/otsmtriz/Home

Osher Digital. (2024). The future of work - How automation is reshaping job roles. https://osher.com.au/blog/future-of-work-automation-reshaping-job-roles/

PODS. (2024). PODS 7.03 Update - From linear referencing to geospatial advantage. https://pods.org/pods-7-03-update-from-linear-referencing-to-geospatial-advantage/

Reason Town. (2024). What is machine learning and deep learning? https://reason.town/what-is-machine-learning-and-deep-learning/

Rufus, V. (2024). Glossary of all the Generative AI terms you will come across. https://vincirufus.dev/posts/glossary-of-generative-ai-terms/

Searchlight.ai. (2024). Webinar: Revolutionizing HR and recruiting with AI. https://www.searchlight.ai/webinar/webinar-revolutionizing-hr-and-recruiting-with-ai

Social Good AI. (2024). Top artificial intelligence books for learning and understanding AI. https://aiforsocialgood.ca/blog/10-must-read-books-on-artificial-intelligence-for-beginners-and-enthusiasts

The Role of Machine Learning in Natural Language Understanding. (2024). *International Journal of Research in Applied Science and Engineering Technology.* https://www.ijraset.com/research-paper/role-of-machine-learning-in-natural-language-understanding

APPENDIX G: TRADEMARK AND LEGAL NOTICES

Trademark Acknowledgments

This book references various trademarks and registered trademarks, each of which is owned by its respective entity. The inclusion of these trademarks is purely for editorial purposes and is not intended to infringe upon any trademark rights.

AI Models and Companies

ChatGPT® is a registered trademark of OpenAI, LP

Claude® is a trademark of Anthropic, Inc.

Anthropic™ is a trademark of Anthropic, Inc.

OpenAI® is a registered trademark of OpenAI, LP

Meta® is a registered trademark of Meta Platforms, Inc. (referenced in the context of Llama)

Technology Companies

Google® is a registered trademark of Google, Inc.

GitHub® is a registered trademark of GitHub, Inc.

X® is a registered trademark of X Corp.

Software and Services

Grammarly® is a registered trademark of Grammarly, Inc.

Legal Notices

This appendix is not intended as legal advice. For specific guidance related to trademarks and other legal issues, readers are encouraged to seek the counsel of qualified legal professionals.

Disclaimer of Warranties

This book is offered without any warranties, either explicit or implied. This includes, but is not limited to, any implied guarantees regarding its merchantability, suitability for a specific purpose, or non-infringement.

Limitation of Liability

The author and publisher cannot be held responsible for any damages that may arise from the use or inability to use the information contained in this book. This includes, but is not limited to, loss of business profits, interruptions, loss of business information, or any other financial losses, regardless of whether they were advised of the potential for such damages.

AI-Generated Content Notice

Portions of this book were developed with the assistance of artificial intelligence language models. The author has reviewed, edited, and curated all AI-generated content in depth. However, the nature of AI technology means that some content may contain unintended inaccuracies or biases.

AI Regulations

Readers should be aware that AI regulations vary by jurisdiction and are rapidly evolving. Implementation of any techniques described in this book should comply with current applicable laws and regulations in your jurisdiction.

Updates and Errata

For updates, corrections, and errata related to this book, please visit:

wrale.com/books

Contact Information

For legal inquiries regarding trademark usage or other legal matters discussed in this appendix, please contact:

Joshua M. Dotson contact@jmdots.com

ABOUT THE AUTHOR

Joshua M. Dotson is a software architect and AI ethics advocate with over 20 years of experience in technology. His current work focuses on creating practical frameworks for human-AI collaboration while maintaining strong ethical principles. Through hands-on experience with AI systems, he has developed unique insights into directing AI effectively and responsibly.

Joshua is particularly passionate about helping professionals harness AI capabilities in ethical and productive ways. His approach combines technical expertise with a deep understanding of how humans and AI can best work together. He regularly explores new ways to make AI collaboration more accessible, ethical, and effective for professionals across industries.

Email: contact@jmdots.com

LinkedIn: linkedin.com/in/jmdots

For updates, resources, and new content, visit: wrale.com/books

Made in the USA
Las Vegas, NV
16 March 2026

43721762R00187